D1259348

Marine Plankton
Life Cycle Strategies

Editors

Karen A. Steidinger, Ph.D.
Chief, Bureau of Marine Research
Florida Department of Natural Resources
St. Petersburg, Florida

Linda M. Walker, M.A.
Biological Consultant
Bureau of Marine Research
Florida Department of Natural Resources
St. Petersburg, Florida

CRC Press, Inc.
Boca Raton, Florida

MIDDLEBURY COLLEGE LIBRARY

ABB-6280

12/1987
Beal

Science
Center
QH
91.8
P5
M36
1984

Library of Congress Cataloging in Publication Data

Main entry under title:

Marine plankton life cycle strategies.

Bibliography: p.
Includes index.
1. Marine plankton. I. Steidinger, Karen A.
II. Walker, Linda M.
QH91.8.P5M36   1984   574.92   83-27218
ISBN 0-8493-5222-3

This book represents information obtained from authentic and highly regarded sources. Reprinted material is quoted with permission, and sources are indicated. A wide variety of references are listed. Every reasonable effort has been made to give reliable data and information, but the author and the publisher cannot assume responsibility for the validity of all materials or for the consequences of their use.

All rights reserved. This book, or any parts thereof, may not be reproduced in any form without written consent from the publisher.

Direct all inquiries to CRC Press, Inc., 2000 Corporate Blvd., N.W., Boca Raton, Florida, 33431.

© 1984 by CRC Press, Inc.
Second Printing, 1986

International Standard Book Number 0-8493-5222-3

Library of Congress Card Number 83-27218
Printed in the United States

INTRODUCTION

The term "plankton" was first applied by Hensen[1] in 1887 for those plant and animal groups of less than 2 cm that inhabit the water column of seas as floaters or weak swimmers. The word is of Greek derivation meaning "drifter"; it infers that these organisms are passive in the aquatic realm. Strickland,[2] however, defined plankton as "organisms suspended in water, without mobility or with very limited mobility which cannot or do not maintain their distribution against effects of local water movements". Many plankton organisms, both plant and animal, can regulate their vertical position and can overcome vertical barriers such as thermoclines or pycnoclines. Within the plankton, there are several widely-used categories such as:

1. phytoplankton vs. zooplankton,
2. auto- or auxotrophic vs. heterotrophic and holozoic,
3. size categories, e.g., picoplankton ($<2\mu$m), ultraplankton (2 to 5μm), nannoplankton (5 to 20μm or 20 μm), microplankton (20μm to 2 mm), macroplankton (2mm to 2cm); "micronecton" (>2cm),
4. residency in the pelagic realm, e.g., holoplankton (complete life cycle in the water column), hypoplankton (part-time pelagic residents; diurnal swimmers from benthic habitats), meroplankton (temporary residents; certain developmental stages separated into pelagic and benthic components), and tychoplankton (swept off the bottom by disturbances such as currents).

These categories provide useful, convenient groupings within the broad category of plankton for purposes of discussion and comparison. However, there is considerable overlap between categories and one should be wary of strict application of these terms. For example, dinoflagellates are considered phytoplankton by botanists and zooplankton by zoologists. Within the dinoflagellate division there are photosynthetic, autotrophic species as well as nonphotosynthetic, heterotrophic species. Copepods are considered holoplanktonic, yet some species have benthic, dormant eggs and these species could be considered meroplanktonic.

Strict adherence to these categories could bias a researcher's experimental design, data analyses and interpretation. Another source of potential bias in the study of marine plankton is the experimental design. Experimental design, including sampling gear and laboratory vs. field observations, influence available data bases and subsequently, conclusions based on these data. Observations and conclusions drawn from laboratory studies and applied to field conditions must be considered cautiously. Laboratory conditions may be optimal and/or unrealistic, and resulting data, therefore, can be biased. In field studies, the choice of sampling gear and time and location of sampling can influence results. Many zooplankters are not passive; they seek specific depths, form schools and respond as schooling animals, regulate buoyancy, and actively seek or trap prey. Omori and Hamner[3] pointed out that commonly used and standardized collection methods mask any heterogeneous distribution. Moreover, developmental stages can have different traits or characters that change in time and space. For example, developmental stages of temperate and boreal copepods live at different depths throughout development. Surface sampling would capture only early and late development stages and would obscure the seasonal migration in these species.

Survival of a species is an adaptive process with variable selection pressures operating at different stages of development. Survival means reproductive success at a level sufficient to provide future generations; it represents "fitness." Stearns[4] defined fitness as "Something everyone understands but no one can define properly," yet he proposed a definition for species fitness as " . . . better represented in future generations than their relatively unfit competitors." In the marine environment, individual survival is dependent on food availability, predation pressure or predator avoidance and competitive advantages (e.g., physiological tolerances and rates — osmoregulation, assimilation efficiency) which can regulate growth and reproductive effort as well as reproductive and dispersal strategies. Survival is an interactive process of individuals with abiotic and biotic factors. The expression of survival is three dimensional in time and space and stimuli very likely act synergistically. Allan[5] expressed this interaction(s) as tradeoffs of resource allocation between reproductive potential, predator avoidance, and competitive advantages.

Margalef[6] stated, "Water movement controls phytoplankton communities," and in a dynamic marine environment this is certainly true; however, genetic variability, life cycles, and behavioral strategies also influence survival and dispersal. Adaptations are diverse in both plant and animal plankton; they include dormancy, size, mobility (motility), morphology, vertical distribution, physiological efficiencies and growth, alteration of generations or clonal propagation, brood size, biological rhythms, feeding mechanisms, and other adaptive mechanisms.

This book was intended to summarize selected plankton life cycles to emphasize the significance of life cycles and their stages in ecological analyses of populations and communities. Without this specific knowledge it is impossible to evaluate occurrence, abundance, distribution, and potential dispersal and extended distribution of many plankters.

Diatoms occur worldwide but are characteristic of high latitude regions, temperate coastal areas, and upwelling zones. Diatoms have relatively simple life cycles that nevertheless show adaptations for survival and dispersal in these fluctuating environments. Vegetative cells have morphological and physiological adaptations to reduce sinking, survive freezing in ice floes, and withstand darkness and unfavorable temperatures. Some species, particularly those in temperate areas, form asexual resting spores or resting cells during environmental stress. These dormant stages may aid population survival but may be more important in species occurrence, species succession, and distribution patterns. Sexual resting stages are not known in diatoms.

Unlike diatoms, the sexual phase in a number of dinoflagellates includes a resting, zygotic cyst. This dormant stage is important in long-term population survival and dispersal. The dormant zygotes require a species specific dormancy period before excystment can occur. Many dinoflagellates also have the ability to form temporary resting stages. These seem to be important in surviving short-term stress such as unfavorable temperature or salinity.

The understanding of sexual life cycles in dinoflagellates has led to important insights into distribution patterns, dispersal, seasonal occurrence, succession and bloom phenomena. Dense blooms of dinoflagellates have been considered evidence of aberrant conditions in nature which allowed the formation of blooms. Study of the life cycles of bloom forming dinoflagellates suggests that these dinoflagellates may be well adapted at synchronizing life history stages with physiochemical changes and patterns in their environment.

The life cycles of planktonic foraminifera and radiolaria are not well known and, as Anderson points out in Chapter 3, are " . . . restricted largely to an understanding of events during maturation, cellular differentiation prior to reproduction, and release of reproductive swarmers." Symbiotic associations between forams and dinoflagellates may be an adaptive tactic to provide energy for certain physiological activities as is true for zooxanthellae and corals. As in forams, or at least those with symbionts, symbionts of radiolaria may also provide energy by supplying photosynthates. The timing and events of swarmer production in both these groups can be influenced by endogenous and exogenous factors. Availability and suitability of food affects the duration of the maturation phase as well as individual viability.

Prior to swarmer production, planktonic forams and radiolaria lose their buoyancy and sink to deeper layers, 50 to 200m. Anderson outlines the adaptation which enhances sinking of the reproductive stage. Several advantages to swarmer release at depth are speculated: (1) introduction to an area of abundant food items, (2) removed from areas of high predator abundance, and/or (3) introduction into boundary layers that aid in dispersal and possible mixing of gametes and variation in the gene pool.

Planktonic copepods are diverse in reproductive tactics, feeding mechanisms, and habitat. Almost all species reproduce sexually and some produce resting eggs or stages that "overwinter." Some resting eggs, called diapausal eggs, require a dormancy period. Resting eggs are benthic and this strategy, or tactic, has adaptive advantages for species survival as do any benthic resting stages of planktonic populations. These eggs can hatch and repopulate the pelagic zone during favorable conditions for ensured growth and survival if phased with primary production pulses in temperate or colder waters. As in phytoplankton, the alternation of benthic-pelagic stages can influence apparent occurrence, abundance, seasonality, and succession. Another "overwintering" mechanism in *Calanus finmarchicus* and other cold water species involves delayed development of late stage copepodites. These copepodites overwinter in deep water, relying on stored wax reserves. In spring, they move into upper water layers, mature to adults and breed.

Many planktonic copepods occupy specific depths or migrate vertically. They, in fact, are not passive "drifters" and can regulate their spatial and temperal distribution. Their migratory behavior can remove them from higher concentrations of predators, introduce them into high phytoplankton layers, put them in contact with potential mates, or put their eggs into moving surface waters.

The study of meroplanktonic larvae offers an instructive contrast to the life cycles of holoplanktonic organisms. By definition meroplankters spend only part of their time in the plankton. The planktonic larvae of benthic invertebrates comprise a major portion of meroplankton. In their larval stages these invertebrates are subject to an entirely different set of environmental variables than in their adult benthic phase. The influence each phase has on the survival and dispersal of the other has been the subject of increasing study. Day and McEdward, in Chapter 5, emphasize the importance of this property of integration of the planktonic and benthic phases in understanding the life cycle of benthic invertebrates.

The variety of larval adaptations for survival and dispersal reflects the diversity of invertebrates with meroplanktonic larvae. However, the study of larval energetics and metamorphosis provides a common basis for interpretation of the diverse patterns of development. Larval nutrition, the length of the planktonic phase, and the presence or absence of suitable substrate affect the success of metamorphosis. The timing and success of metamorphosis in turn affects the success of distribution and survival of the adult population.

Fish have evolved a variety of reproductive tactics to insure survival, including protection of young and dispersal mechanisms. Survival of planktonic fish larvae depends on many integrated factors: hatching time and place of eggs, temperature, food availability and suitability, and transport of larvae, often to nursery grounds. In those species with yolk sacs, larvae must feed and assimilate within a short period after yolk sac absorption or die. This is called the critical period and can determine recruitment success and year class strength. Various authors have pointed out that predation of fish larvae is high and is inversely related to growth rate of larvae. Therefore larval survival is ultimately keyed to food supply. Fish larvae feed on other plankton (usually 40 to 100 μm) such as copepod nauplii, copepodites, mollusk larvae, and phytoplankton.

Oceanic processes such as currents, frontal systems, and convergences can entrain fish larvae as well as other plankton. Such boundary layers or water masses also contain suitable and abundant prey for fish larvae. Several studies have documented that fish eggs and larvae are associated with high chlorophyll patches, for example, in upwelled water masses or meanders. Copepod nauplii and copepodites are associated with these high chlorophyll patches or layers as well. Spawning time and area appear to be critical to the potential success of developing larvae by placing them in areas of high density food, or plankton, patches. The synchrony of spawning seasons with zooplankton pulses may insure growth and viability. These same water masses or boundary layers may also insure survival through reduction of advective loss of eggs and larvae, transport of larvae to nursery grounds, and maintenance of larvae in areas suitable for growth.

The authors, in their respective chapters, have emphasized adaptations for reproductive success, development and protection of early life cycle stages, and dispersal mechanisms. Most discussions address natural events; however, man's influence in species distribution and success should not be ignored. The effects of pollution, overfishing, and alteration or destruction of habitat are increasingly evident, especially in food organisms important to man such as oysters, sturgeon, and whales. Man can also, intentionally or unintentionally, introduce new species to an area which may or may not disrupt the native community. Study of life histories of marine plankters includes members of a large portion of the marine food web from primary producers to benthic and pelagic predators. Knowledge of the basics as well as the interrelationships of marine plankton life histories can be applied in fisheries management, red tide prediction, coastal zone management, and habitat restoration and enhancement.

K. A. Steidinger
L. M. Walker
August, 1983

REFERENCES

1. Hensen, V., Über die Bestimmung des Planktons oder des im Meere treibenden Materials an Pflanzen und Thieren, *Konn. Wiss. Untersuch. Deutsch. Meere*, 5, 1, 1887.
2. Strickland, J. D. H., Measuring the production of marine phytoplankton, *Fish. Res. Board Can. Bull.*, 122, 1, 1960.
3. Omori, M. and Hamner, W. M., Patchy distribution of zooplankton: behavior, population assessment and sampling problems, *Mar. Biol.*, 72, 193, 1982.
4. Stearns, S. C., Life history tactics: a review of the ideas, *Q. Rev. Biol.*, 51, 3, 1976.
5. Allan, J. D., Life history patterns in zooplankton, *Am. Nat.*, 110, 1, 1976.
6. Margalef, R., Life-forms of phytoplankton as survival alternatives in an unstable environment, *Oceanol. Acta*, 1, 493, 1978.

THE EDITORS

Karen A. Steidinger, Ph.D. is Chief, Bureau of Marine Research, Florida Department of Natural Resources, St. Petersburg, Florida.

Dr. Steidinger graduated from the University of South Florida with a B.A. degree in Zoology in 1968, an M.A. in Marine Science in 1971, and a Ph.D. in Biology in 1979.

Dr. Steidinger is a member of the American Association for the Advancement of Science, American Association of Stratigraphic Palynologists, American Microscopical Society, American Society of Limnology and Oceanography, Biological Society of Washington, Florida Academy of Sciences, International Phycological Society, Fellow of the American Institute of Fishery Research Biologists, and is an American Fisheries Society Certified Fisheries Scientist. Among a variety of advisory groups, she serves on the Gulf of Mexico Fishery Management Council Standing Scientific and Statistical Committee and the Florida Institute of Oceanography Advisory Board. In 1983, she received the Florida Academy of Sciences Medalist of the Year Award.

Dr. Steidinger has organized or chaired conferences and sessions at the international and national level, presented numerous invited lectures on phytoplankton and red tides, and published over 30 scientific papers. Recently, she was an invited instructor at the University of Oslo's Third International Phytoplankton Course. Dr. Steidinger's research interests are the systematics and ecology of marine dinoflagellates, particularly toxic species impacting fisheries.

Linda M. Walker, M.A. is a biological consultant at the Bureau of Marine Research, Florida Department of Natural Resources, St. Petersburg, Florida.

Ms. Walker graduated in 1974 from the University of South Florida with a B.A. degree in zoology and in 1978 received an M.A. in Marine Biology from the same university.

Ms. Walker is a member of the American Society of Zoologists, American Society of Limnology and Oceanography, Biological Association of the United Kingdom, Biological Society of Washington, the Crustacean Society, and Sigma Xi.

Ms. Walker has authored and co-authored research papers on marine plankton, both phyto- and zooplankton. Her current research interests include copepod development and taxonomy and larval fish feeding dynamics.

CONTRIBUTORS

O. Roger Anderson, Ph.D.
Professor of Natural Science and Senior
  Research Associate
Marine Biology
Columbia University
Palisades, New York

Charles C. Davis, Ph.D.
Professor of Biology (Ret.)
Memorial University of Newfoundland
St. John's, Newfoundland, Canada

Randy Day, Ph.D.
Associate Professor of Biology
Brigham Young University—
  Hawaii Campus
Laie, Hawaii

David L. Garrison, Ph.D.
Assistant Research Marine Biologist
Center for Marine Studies
University of California
Santa Cruz, California

Mark M. Leiby
Senior Ichthyologist
Bureau of Marine Research
Florida Department of Natural
  Resources
St. Petersburg, Florida

Larry R. McEdward
Graduate Student, Dept. of Zoology
  and Friday Harbor Laboratories
University of Washington
Seattle, Washington

ACKNOWLEDGMENTS

We would like to thank the following reviewers for their comments and criticisms: Drs. Marsh Youngbluth, Greta Fryxell, and Wayne Bock. Additionally we thank all authors for their contributions, cooperation, and patience, and Patricia Boyett for her typing and assistance.

TABLE OF CONTENTS

Chapter 1

PLANKTONIC DIATOMS

David L. Garrison

TABLE OF CONTENTS

I. INTRODUCTION

Diatoms are important components of marine and freshwater phytoplankton communities. They are usually associated with high latitude regions, temperate coastal areas, and upwelling systems.[1-6] These environments are all characterized by dramatic temporal fluctuations, and we would expect diatom life history traits to reflect adaptations for survival in these fluctuating environments. Diatoms are perhaps the best known of all planktonic algal groups,[3-10] but the study of the ecological significance of their life cycles has barely begun.[5] In this chapter, I will attempt to synthesize what information we do have. While I have made no particular effort to limit consideration to marine planktonic diatoms, the emphasis on marine systems will be apparent.

Diatoms have some unique characteristics that are important to understand before we can examine the significance of their life cycle features. Unlike most algae, diatoms have a rigid silicate frustule encasing the vegetative protoplast (Figures 1 and 2). Traditionally, shape and symmetry have been the basis for separating diatoms into two orders. Centric diatoms (Centrales) have essentially circular valves and are radially symmetrical about the pervalvar axis (see Figure 1a), whereas pennate diatoms (Pennales) have elongated or ''boat-shaped'' valves that are bilaterally symmetrical with reference to the apical plane (see Figure 1b). Although these traditional orders may not accurately reflect evolutionary relationships,[13] they may have some ecological significance: centric diatoms are usually planktonic, and pennates are characteristically benthic.

Unlike the vegetative cells of other planktonic algae, those of diatoms do not possess flagella. They are therefore nonmotile (as planktonic forms) and sink in nonturbulent surroundings. Sinking rates are related to cell size and shape, colony size, and physiological condition. In general, sinking rates increase with cell and colony size and are greater in senescent populations.[14]

Diatom life cycles include vegetative, sexual, and resting stages (Figures 2 and 3). Normally, diatoms reproduce by vegetative division, and for many species the vegetative stage is the only one commonly observed. Although vegetative cells are capable of existing independently, they often remain attached after cell division and form distinct colonies. During cell division, a new half of each sibling theca is formed inside the theca of the parent cell, resulting in a progressive decrease in cell size with repeated cell divisions (Figure 4). Most diatoms undergo this size reduction, and to restore maximum cell size they must occasionally interrupt vegetative growth with a sexual cycle which restores maximum cell size (see Figure 3) . The restoration of cell size by sexual reproduction is a unique feature of diatoms.[16] Centric and pennate diatoms differ somewhat in the details of sexual reproduction, but in both groups, zygote formation is followed by the formation of an enlarged cell called an auxospore, which then develops into a vegetative cell whose dimensions are near the maximum size characteristic for the species.

Many diatoms form resting spores or resting cells (Table 1). Resting spores, common in centric diatoms[15,18-29,31,33-35,37] and rare in pennates,[30,36] are heavy-walled stages that are morphologically distinct from vegetative cells (Figure 2 c,d,f). Resting cells, on the other hand, are similar to vegetative cells, except for altered cytoplasmic characteristics — for example, darker pigmentation or unusual plastid structure.[31-34] Resting spores are formed from vegetative cells, but in some species they are formed from auxospores after sexual reproduction.[34,38-41]

II. THE SIGNIFICANCE OF LIFE CYCLE STAGES

Diatoms are associated with highly seasonal environments, and one might therefore

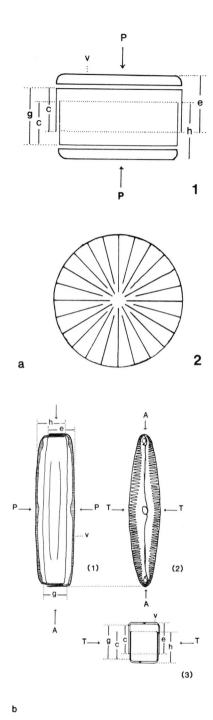

FIGURE 1. Diagrammatic representation of a diatom frustule. (a) Centric type diatom: (1) girdle view (2) valve view. Axes: P, prevalvar. Structure: e, epitheca; h, hypotheca; v, valve; c, cingulum; g, girdle. (b) Pennate type diatom: (1) broad girdle view (2) valve view (3) narrow girdle view. Axes: A, apical; P, perval-var; T, transapical. Structure: e, epitheca; h, hypotheca; c, cingulum; g, girdle; v, valve. Frustule = epitheca + hypotheca; theca = valve + cingulum (terminology follows that in Anon;[11] adapted from Cupp.[12] With permission.)

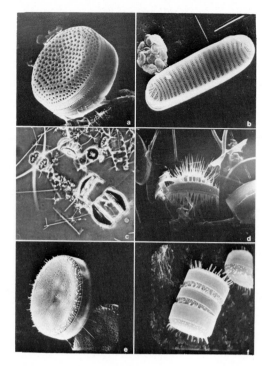

FIGURE 2.   (a) *Thalassiosira trifluta*, centric diatom. (b) *Nitzschia curta*, pennate diatom. (c,d) *Chaetoceros* resting spores. (e) Vegetative cell of *Thalassiosira antarctica*. (f) Resting spore of *T. antarctica*.

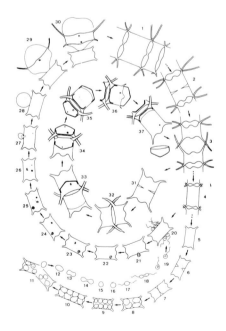

FIGURE 3.   Life cycle stages in *Chaetoceros didymum:* vegetative cell divisions 1 to 4; sexual cycle 6 to 30, auxospore formation 27 to 30; resting spore cycle 31 to 37. (After von Stosch et al.[15] With permission.)

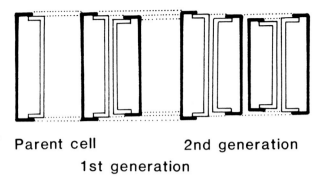

Parent cell   2nd generation
1st generation

FIGURE 4. Diagrammatic representation of a diatom in girdle view showing the progressive decrease in size with cell divisions: the MacDonald-Pfitzer rule. (After Hendey.[13] With permission.)

Table 1
SUMMARY OF SPECIES FORMING RESTING STAGES[a]

Distribution taxon	Number species	Ref.	Distribution taxon	Number species	Ref.
Planktonic diatoms			Planktonic diatoms (continued)		
Thalassiosira	8+		*C. lauderi*		23
T. nordenskioeldii		26,29	*C. didymum*		15,23
T. gravida		28	*C. diadema*		21,23
T. constricta		28,29	*C. curvisetum*		23
T. scotia		18	*C. debile*		23
T. antarctica		20,26,28,29	*C. radicans*		35
T. fallax		28	*C. socialis*		23
T. australis		26	*Ditylum brightwellii*		
Bacteriosira fragilis		24	*D. sol*		30
Detonula confervacae		22,26	*Odontella weissfolgii*		25
Porosira glacialis		24	*O. litigosa*		25
Skeletonema costatum	?	33	*Fragilaria*	?	20
Hemiaulus	?	13	*Nitzschia grunowii*		34
Rhizosolenia styliformis		30	*Achnanthes*	?	20
R. setigera		22	*Amphora coffaeformis*		32
R. eriensis[b]		22	*Melosira italica*[b]		31
R. alata			subsp. *subarctica*		
Ceratulina	2	37	*M. granulata*[b]		31
Eucampia balaustium		25	*M. ambigua*[b]		31
Leptocylindrus danicus		22			
Stephanopyxis palmeriana			Nonplanktonic diatoms		
S. turris		22,27	*Diatoma anceps*		30
Acanthoceros	?	20	*Meridon circulare*		30
Bacteriastrum hyalinum		19	*Eunotia*	3	
B. delicatulum		22	*E. soleirolii*		30
Chaetoceros	39+		*Navicula grevilleana*		13
C. teres		23	*N. cuspidata*		36

[a]   Diatoms are reported to form resting stages.[12,13,17,20,24,34] The number of species forming resting stages is summarized for genera with many spore-forming species. Only recent reference sources are cited in table.

  +, the number of species is probably greater.
  ?, uncertain.

[b]   Freshwater species.

expect to find specialized life cycle stages for maintaining populations during unfavorable conditions. Diatoms, however, have a limited array of life cycle stages, and resting spores and resting cells are the only ones that appear to be specialized for survival or dispersal. Auxospores, which are formed during the sexual cycle, have not been considered as survival stages because they develop immediately into vegetative cells.[34] Microspores, which are apparently spermatogonia,[13,16] have been reported to give rise to vegetative populations in some *Chaetoceros* species,[42-44] but these reports have not been well supported. Our present evidence suggests that many species do not have resting stages and must therefore be capable of surviving during unfavorable periods either as undifferentiated vegetative cells, or as cells with cytological changes that are not accompanied by alterations in the frustule. Such forms are not easily perceived as survival stages, but they may well function in that manner.

A. Survival of Vegetative Populations

The results of survival experiments on several species indicated they could survive unfavorable winter conditions of low light and low temperatures (Table 2). Many of the species tested have no recognized resting stages, yet are capable of long term survival. For example, Smayda and Mitchell-Innes[45] reported dark survival by *Skeletonema costatum* of up to 24 weeks at incubation temperatures of 2°C, and Umebayashi[46] reported survival times up to nine months at 5°C for *Skeletonema* when the cultures were regularly exposed to low light levels. Only a few experiments, however, have tested survival in warm, low nutrient conditions, such as those prevalent during the summer diatom minima in temperate waters (see Table 2). Dodson and Thomas[44] reported approximately eleven weeks survival under these conditions. Antia and Cheng[50] tested dark survival at warm temperatures, but a combination of warm temperatures and high light levels would be more representative of natural conditions.

Winter conditions in polar regions are unfavorable for algal growth, and Rhode[52] proposed that heterotrophic metabolism was necessary to explain survival during the prolonged winter darkness. Bunt and Lee,[51] however, demonstrated that ice algal species were able to remain viable for three months in darkness (see Table 2) and that heterotrophic nutrition was not needed. Hellebust and Lewin[53] concluded that, while heterotrophic nutrition could increase survival time, it was not necessary for dark survival; moreover, they found that heterotrophy was rare in planktonic forms. In the Weddell Sea, planktonic diatoms are maintained in pack ice floes for a season or longer[54] and may serve as a seed population for the pelagic system when seasonal melting releases them. Shandelmeier and Alexander[55] have shown that such seeding takes place in an arctic ice-edge environment. Thus, diatoms not only survive winter conditions in polar regions, but contrary to what one might expect, resting stages are not prevalent.[34,56,57]

Oceanic diatoms could escape unfavorable surface water conditions during the summer by sinking into deeper water; the recovery of viable cells from deep water offers some evidence for such a survival strategy.[58-64] The ecological significance of some of these deep algal populations is questionable, however, because they are located below the depth of seasonal mixing, and it is not clear how these populations would return to surface waters. Moreover, Silver and Alldredge[64] found that the deepwater forms that they examined were a unique collection of survivors and not representative of surface populations. Venrick has proposed the hypothesis that populations in the chlorophyll maximum layer near the bottom of the euphotic zone represent a senescent survival stage in oceanic population cycles.[60,62] Her model is worth considering because these populations are within a depth range from which they could be resuspended during mixing events.

Table 2
SELECTED DATA FROM SURVIVAL
EXPERIMENTS ON VEGETATIVE CELLS

Species	Experimental conditions[a] (°C)	Survival weeks	Source
Skeletonema costatum	20 — D	1	50
	20 — D	4	49
	∿18 — D	7	48
	15 — D	7	45
(winter clone)	15 — D	14	33
(autumn clone)	15 — D	11	33
(pacific clone)	15 — D	16	33
	13 — D	8	47
	5 — D	16	46
	5 — D	36	46
	2 — D	20	33
	∿2 — D	24	49
Thalassiosira fluviatilis	20 — D	8+	50
	15 — D	19	33
	20 — D	7	50
Cyclotella nana	5 — D	16	46
(= *T. pseudonana*)	5 — I	36	46
Phaeodactylum tricornutum	20 — D	24+	50
	5 — D	24	46
	5 — I	136	46
Chaetoceros gracile	20 — D	8+	46
C. didymum	15 — D	13+	45
C. curvisetum	15 — D	13+	45
C. sociale f. *radians*	15 — D	2	33
(= *C. radians*)			
C. calcitrans f. *pumilus*	5 — D	16	46
	5 — D	36	46
C. affine	∿20 — O	∿11	44
C. fragile	−1.8 — D	12+	51
Fragilaria sublinearis	−1.8 — D	12+	49
= *Nitzschia sublineata*			

Note: (Compiled from Matsue,[47] Takano,[48] Antia (unpublished) cited by Smayda and Mitchell-Innes,[49] Antia and Cheng,[50] Umebayashi,[46] Bunt and Lee,[51] Smayda and Mitchell-Innes,[45] Hargraves and French,[33] Dodson and Thomas.[44])

[a] D, dark incubations.
I, exposed to intermittant low light levels.
O, oligotrophic conditions, 14/10 light/dark incubation, temp. ∿20°C.
+ survival for maximum duration of experiment.

Although it is often assumed that resting stages are necessary for the survival of coastal species, many coastal species have no recognized resting stage[65] and apparently persist as vegetative populations. Wood,[66] Braarud and Steeman-Nielsen,[8] and Hargraves and French[33] have suggested that *Skeletonema costatum* can survive as a benthic population in shallow water sediments. Mixing events in coastal waters can resuspend sinking phytoplankton populations,[67-69] but the importance of these resuspended cells as a seed stock for phytoplankton populations has not been extensively examined.[70]

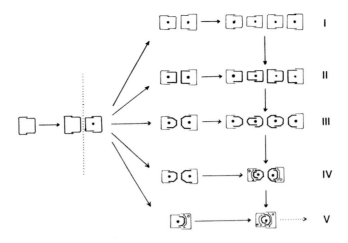

FIGURE 5.   Model for the morphological relationships between vegetative cells and resting stages. (I) Normal vegetative division; (II) resting cell development; (III) exogenous spore development; (IV) semi-endogenous spore development; (V) endogenous spore development. Partially formed spores in II and III may or may not continue development. (Modified from Syversten.[26] With permission.)

B. The Role of Resting Stages
1. Resting Stage Formation

The process by which resting stages are formed may not differ markedly from vegetative cell division. For example, mitosis is required to form specialized resting spore valves as it is for vegetative valves.[15,71] Syversten[26] presented a model to show the morphological relationship between vegetative cells and the different types of resting stages that have been described (Figure 5). During resting spore formation, cell division eventually terminates with nonviable or residual cells (Figure 5, III to V); therefore, a variable number of resting spores are formed from one vegetative cell. Fryxell et al.[20] reported that a variable number of spores were formed by *Thalassiosira antarctica*, and they suggested that the number of spores formed may be limited by some material in the cell rather than under strict genetic control. This observation suggests that the final relationship between spores and vegetative cells may not be a fundamental species specific characteristic. Whereas resting stages with specialized frustules may require a series of cell divisions to form, it is not clear that resting cells form in this manner. For example, Anderson[32] reported that resting cells of *Amphora coffaeformis* developed over a four week period in culture, but he did not indicate if cell divisions accompanied the cytoplasmic reorganization of these resting cells.

2. Adaptations for Survival

Recent experimental evidence supports the role of resting spores in vitro. Hargraves and French[33] initially established that dark survival of spores was greater than that of vegetative cells. In subsequent studies, survival times of up to 22 months have been reported (Table 3), but survival under natural conditions may not be as long, because temperature and other environmental conditions have a pronounced effect. For example, Durbin[72] demonstrated that resting spores of *Thalassiosira nordenskioeldii* and *Detonula confervacea* could not survive summer temperatures ($\sim 20°C$) in shallow areas of Naragansett Bay and concluded that the seasonal appearance of these boreal species could not be attributed to local resting stages. Davis et al.[40] and Hollibaugh et al.[73] observed that resting spores did not survive in sediments where there were high tem-

Table 3
SURVIVAL CHARACTERISTICS OF DIATOM RESTING
SPORES IN DARK SURVIVAL EXPERIMENTS

A. Maximum survival times

Species	Temp. (C)	Survival[a] (days)	Source
Chaetoceros diadema	2	140+	33
	2	645	70
C. vanheurckii	2	645	70
C. didymum	2	645	70
C. sociale f. radicans	15	140+	33
Detonula confervacea	0	576	69
	2	140+	33
Eunotia soleirolii	5	400	30
	15	400	30
Leptocylindricus danicus	3	214+	38
	10	214+	38
	10	401	34
Stephanopyxis turris	15	112	33
Thalassiosira nordenskioeldii	0	576	69

B. Survival time (days) as a function of temperature

| Species | Temp. (°C) | | | | | | Source |
	0	∿2	5	10	15	20	
Detonula confervacea	576		220	85	42	<7	69
T. nordenskioeldii	576		220	105	42	<7	69
L. danicus				401	191	217	34
		214+		214+			38
E. soleirolii	58	58	400		400		30

Note: Compiled from von Stosch and Fecher,[30] Hargraves and French,[33] Hargraves and French,[34] Davis et al.,[40] Durbin,[72] Hollibaugh et al.[73]

[a]   +, spores were still viable at maximum duration of experiment.

peratures and a reducing environment. Moreover, maximum survival times may overestimate the survival potential of resting spores, since in these studies the number of viable spores decreased rapidly with time, and only a few viable spores remained at the maximum survival time.[40,73] There is less information about the survival capabilities of resting cells than there is about resting spores. Lund[31] reported that resting cells of *Melosira italica* subsp. *subarctica* could survive in dark, anaerobic conditions for up to three years. Experiments of shorter duration suggested similar abilities for *M. granulata* and *M. ambigua*. Anderson[32] reported dark survival of resting cells of *Amphora coffaeformis* for at least two months at 7°C.

There appear to be physiological and cytological adaptations that account for extended survival capabilities of resting stages. French and Hargraves[74] showed that resting spores have a higher carbon/nitrogen ratio, higher concentrations of carbon and chlorophyll, and lower dark respiration rates in comparison with vegetative cells. These characteristics may be established in resting spores during spore formation. For example, Fryxell et al.[20] reported that there was an unequal distribution of cytoplasmic material between the resting spore and residual cell in the last cell division of spore formation in *Thalassiosira antarctica* that resulted in a concentration of chloroplasts in the resting spore and usually none in the residual cell. Doucette[75] observed that this unequal distribution also resulted in fewer mitochondria in the spores than in the residual cell. Resting cells, however, appear to be formed by reorganization of cytoplasm. Anderson[32] reported that reduction of vacuole size, fusion of mitochondria, and increased lipid reserves, accompanied resting cell formation in *Amphora coffaeformis*. In subsequent experiments, Anderson[76] demonstrated that respiration rates were lower in resting cells than vegetative cells. Such evidence is consistent with the idea that these forms function as survival stages, but relatively few species have been examined.

One might also expect resting stages to have morphological adaptations for reducing contact between the cytoplasm and an unfavorable external environment. In diatoms, this reduction may be accomplished by altering structural features of the frustule. For example, Hargraves reported that resting spores in *Chaetoceros* were markedly different from vegetative cells (see Figure 2 c and d), and he suggested that there was restricted contact between the spore plasmalemma and the external environment;[23] however, in other genera (e.g., *Thalassiosira, Detonula, Odontella,* and *Stephanopyxis*) resting spore and vegetative cell valves have similar structural features (see Figure 2 e and f) and contact between the spore and the environment may be less restricted in these genera than in *Chaetoceros*.[22,25,26]

Previously, the potential role of resting stages in overwintering was emphasized,[77] but more recent studies suggest they could function over shorter periods. One reason for suspecting this role is that most resting spores have no obligatory period of dormancy,[34] and are capable of immediate germination when provided with suitable growth conditions.[29,34,43,73,74,78] The freshwater pennate diatom, *Eunotia soleirolii*, requires 4 to 5 weeks of dark incubation before germination is possible and is a clear exception to the pattern of immediate germination;[30] however, *E. soleirolii* is not planktonic and there is no reason its response should necessarily be the same as planktonic forms.

Resting spores may have functions other than benthic survival during unfavorable growth conditions. Hargraves and French[33] have shown that resting spores survive drying better than vegetative cells, but they were unable to find evidence for aerial dispersal. In another study they showed that resting spores of *Chaetoceros sociale* f. *radians* and *Detonula confervacea* survived passage through the gut of a copepod grazer; other species did not show this ability.[34]

3. Resting Stage Cycles in Natural Populations

Resting spores have been regarded as the primary mechanism for maintaining neritic diatom populations. Gran[77] originally proposed that resting spores were benthic resting stages. Essential features of Gran's hypothesis were (1) that resting spores are initiated by poor growth conditions, (2) that spores settle locally and persist through unfavorable periods as benthic resting stages, and (3) that spores are resuspended and serve as a seed population when favorable growth conditions return to the water column. Although Hardy and Gunther[79] and Hart[80,81] suggested that spores could function as pelagic as well as benthic stages, Gran's hypothesis was essentially untested until Hargraves and French[33] examined species cycles in Naragansett Bay and were unable to explain vegetative population cycles by persistence of resting spores in sediments.

I followed resting spore cycles in several *Chaetoceros* species in a shallow coastal environment and concluded that Gran's model may be essentially correct.[82,83] It is difficult to draw firm conclusions from any single study, but most of our current information provides convincing evidence that resting spores function as survival stages in coastal populations.

Resting spore formation in natural populations is often associated with nutrient depletion following a peak in population development[33,40,73,81-88] and often precedes the decline or disappearance of a species from the phytoplankton.[30,33,40,82-86,89] In culture, a variety of factors are known to influence spore formation, but where spore formation has been carefully examined, low nitrogen levels are frequently the necessary and sufficient conditions.[34,40,74]

Resting spores sink after they are formed in the plankton. Sinking rates reported for resting spores range from two meters per day[40,74] to ten meters per day.[14] Spores generally sink faster than vegetative cells;[34,74,90] these differential sinking rates may result in a separation of these stages in the water column,[40] and the faster sinking rates of spores should increase their chances of local settlement.

The fate of resting spores in the environment is not well established. Spores may settle locally or be transported into other shallow or deep water areas. Spores have been recovered from shallow-water sediments following their appearance in planktonic populations,[33,82,83] but the evidence for their persistence in sediments is largely circumstantial. Hargraves and French[33] reported that spores did not persist over time in the shallow sediments of Naragansett Bay. Zgurovskaya,[91] however, recovered spores from sediments throughout the year as I also did.[82,83] My own evidence[82] suggests that spores may be removed from shallow sediments by mixing and transport, but may persist with fine sediments in deeper water.

Spores have traditionally been regarded as benthic stages, but they are occasionally found in oceanic regions where it is unlikely that they could function as benthic stages.[79,81,89] There is little evidence that spores function as pelagic survival or dispersal stages; however, local current patterns could return spores to inshore surface waters. For example, I suggested that such transport is possible in a coastal upwelling systems[82] (Figure 6). Hargraves and French[34] suggest that spores associated with marine snow[94] may persist as pelagic rather than benthic resting stages. However, sinking rates of 50 to 100 m/day have been reported for marine snow;[65,95,96] therefore, this association may actually accelerate spore transport to sediments or deep water.

Dynamic mixing processes which are capable of resuspending spores are well known,[67-70,82] and there is some evidence that seasonal phytoplankton populations develop from resuspended spores. For example, Kashkin[97] recovered and germinated the spores of spring bloom species from sediments long before these species appeared in the plankton. I observed spores in sediment and in the water column before the onset of species blooms, but there were always vegetative cells present and there was no practical way of establishing that these spores were in fact the initial source of developing populations.[82,83]

Resting spores and resting cells may have similar ecological roles as survival stages, but resting cells are poorly known, and we have limited information about their significance. Lund's[31] studies on the population cycles of the freshwater diatom *Melosira italica* subsp. *subarctica*, however, provides the clearest example of the role of resting stages in natural population cycles. Lund observed a reciprocal relationship between vegetative populations of *M. italica* in the water column and resting cells in benthic deposits. Vegetative blooms of this species occurred in the early spring following the resuspension of benthic colonies by mixing processes. When the water column became stratified, vegetative populations immediately disappeared and there was a corresponding increase in benthic populations.

III. DISCUSSION AND SUMMARY

Present evidence suggests that planktonic diatoms have relatively simple life histories with few specialized stages. Many species appear capable of maintaining populations

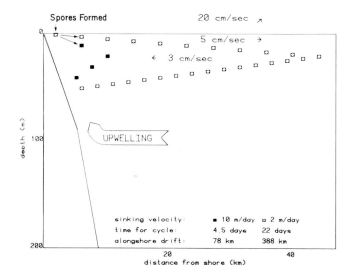

FIGURE 6.   A model for estimating the potential of retaining sinking resting spores in a coastal upwelling circulation cell. (From Garrison[82] with data from: Huyer and Smith,[92] fig. 5; current velocities are from Bryden,[93] figure 3; spore sinking rates are from Davis et al.[40] and Smayda[14].)

by vegetative growth, and some are capable of forming resting stages. Although we lack detailed life history information for most species, it seems unlikely that morphologically distinct stages have been overlooked. On the other hand, dormant stages that possess physiological adaptations not accompanied by changes in frustule morphology (e.g., some resting cells), may indeed have been missed, and generalizations about adaptive strategies may be premature.

Hargraves and French[33,34] suggested that physiological resting stages may in fact be common, particularly among coastal species, but their existence would be difficult to establish because it is not clear what traits would distinguish them. One would expect resting stages to have decreased metabolic activity and more storage products,[30,32,76,98] but these and other attributes of resting cells (e.g., lipid production, darker pigmentation, and condensed cytoplasm) have also been associated with senescence.[98,99] It will first be necessary to ascertain whether these morphological and cytological changes are adaptations for dormancy or simply the characteristics of senescence. It may be necessary to consider when cytological changes are initiated in relation to the rest of the life cycle in order to assess the adaptive value of these changes. Some resting spores form fully after two divisions of a vegetative cell that is committed to form spores (see Figure 5). Resting cells, which do not alter their frustule structure, may or may not require these additional cell divisions, but there should also be selection for initiating cytological adaptations at the onset of stress and forming functional resting stages before viability is reduced.

Although we have little information about resting cells, it seems clear that resting spores are well suited for being survival stages.[34] There has been some question, however, whether resting spores are true dormant stages.[30,34,74] Sussman[100] defines several phenomena related to dormancy:

1.   Dormancy — any rest period or reversible interruption of the phenotypic development of an organism.
2.   Constitutive dormancy — a condition wherein development is delayed due to an innate property of the dormant stage.

3. Exogenous dormancy — a condition wherein development is delayed because of unfavorable chemical or physical conditions of the environment.
4. Maturation — the complex change associated with the development of the resting stage or dormant organism.
5. Activation — the application of the environmental stimuli which induces after-ripening.
6. After-ripening — that part of the dormant period during which the changes occur which lead to germination.
7. Germination — the first appearance of the stage which follows the spore or other propagule.

According to Sussman's definition, diatom resting spores do show dormancy. There are differences between the maturation of resting spores and resting cells. For example, Anderson[33] reported that resting cell formation in *Amphora coffaeformis* required four weeks, whereas resting spores usually form within a day or so. We do not know the ecological significance of this difference.

Consideration of the environments where resting spores are common has given us some clues to their adaptive role. There seems to be a pattern to the geographic distribution of spore-forming species: they are rare in polar regions and common in temperate regions.[34] Hargraves and French[34] suggest that species that form spores are uncommon in polar regions because low nutrients are often the cue for spore formation and nutrients are usually not limiting in polar environments. An alternative explanation is that modifications of the frustule may not be necessary for survival stages in cold water environments: several polar species form resting stages with only minor modifications of the vegetative frustule.[20,25,26] There may also be other morphological or cytological modifications in polar species that have not yet associated with a survival strategy. Because of prolonged winter darkness, one would expect to find dormant stages whose formation is initiated by changing light levels as they are in several other organisms. Dormant stages simply may not have been recognized in pennate diatoms that often dominate polar phytoplankton and ice algal communities.

The prevalence of resting spores among temperate species suggests that highly modified resting stages may be necessary in this environment. Survival may be difficult under warm, low nutrient, and high light conditions that prevail during periods of water column stratification in temperate waters. Hargraves and French[34] suggested that spores provide escape from these conditions. Environmental conditions in temperate regions are heterogeneous in both space and time, and this intrinsic variability may explain the prevalence of resting spores among temperate species.[34] In contrast to a long term survival stage suitable for overwintering, resting spores may have a role as short term stages in temperate regions where stratification is periodically interrupted by mixing events and phytoplankton blooms are repeated events throughout a seasonal cycle.[82,83,101] For example, upwelling regions[86,101,102] and dynamic coastal systems[68-71] are regions where one would expect resting spores to have a role as short-term survival stages. There are intraspecific variations in the ability to form resting spores[15,34,103] that may reflect ecological strains of spore-forming species in environments where resting spores are advantageous.

While it is not apparent that specialized stages such as resting spores are absolutely necessary for insuring population survival, these stages may be important in determining species cycles, succession, and distribution patterns. It seems clear that to understand the dynamics of phytoplankton communities, we will have to pay closer attention to the role of life cycle events. It is also important to realize that our present simplistic view of population cycles, particularly in oceanic populations, may be the result of how little we know about the natural history of diatoms in aquatic environments.

ACKNOWLEDGMENTS

I am grateful to Paul Hargraves and Fred French for providing me with a preprint of their review, *Diatom Resting Spores: Significance and Survival Strategies*. Critical reviews by Greta Fryxell and Mary Silver were helpful in preparing this chapter. I am particularly indebted to Ellen Chu for her editing of the manuscript. Figures were prepared by Shannon Brownlee and James G. Mitchell. Resting spores shown in Figure 2 b and c were provided by J. T. Hollibaugh. Scanning electron micrographs of *Thalassiosira antarctica* in Figure 2 e and f were provided by Greta Fryxell. Parts of this work were supported by N.S.F. DPP 80-20616.

REFERENCES

1. Hutchinson, G. E., *A Treatise on Limnology*, Vol. II, John Wiley and Sons Inc., New York, 1967.
2. Allen, M. B., High-latitude phytoplankton, *Annu. Rev. Ecol. and Syst.*, 2, 261, 1971.
3. Patrick, R., Factors effecting the distribution of diatoms, *Bot. Rev.*, 14, 473, 1948.
4. Patrick, R., Ecology of freshwater diatoms — diatom communities, in *The Biology of Diatoms*, Werner, D., Ed., Univ. of Calif. Press, Berkeley, 1977, chap. 10.
5. Guillard, R. R. L. and Kilham, P., The ecology of marine planktonic diatoms, in *The Biology of Diatoms*, Werner, D., Ed., Univ. of Calif. Press, Berkeley, 1977, chap. 12.
6. Schrader, H.-J. and Schuette, G., Marine diatoms, in *The Sea, Volume 7 The Oceanic Lithosphere*, Emiliani, C., Ed., John Wiley and Sons Inc., New York, 1981, 1179.
7. Lewin, J. C. and Guillard, R. R. L., Diatoms, *Annu. Rev. Microbiol.*, 17, 373, 1963.
8. Oppenheimer, C. H., *Marine Biology II. Proceedings of the Second International Interdisciplinary Conference*, New York Academy of Sciences, New York, 1966.
9. Ettl, H., Müller, D., Neumann, K., and Stosch, H. A. von, Vegetative fortpflanzung, parthenogenese und apogamie bei algen, *Handb. Pflphysiol.*, 18, 597, 1967.
10. Werner, D., *The Biology of Diatoms*, Univ. of Calif. Press, Berkeley, 1977.
11. Anonymous, Proposal for a standardization of diatom terminology and diagnosis, *Nova Hedwigia Beih.*, 53, 323, 1975.
12. Cupp, E. E., Marine planktonic diatoms of the west coast of North America, *Bull. Scripps Inst. Oceanogr.*, 5, 1, 1943.
13. Hendey, N. I., An Introductory Account of the Smaller Algae of the British Coastal Waters, V Bacillariophyceae (Diatoms), Her Majesty's Stationery Office, London, 1964.
14. Smayda, T. J., The suspension and sinking of phytoplankton in the sea, *Oceanogr. Mar. Biol. Annu. Rev.*, 8, 853, 1970.
15. Stosch, H. A. von, Theil, G., and Kowallik, K., Entwicklungsgeschichtliche untersuchungen an zentrischen diatomeen, V. Bau und lebenszyklus von *Chaetoceros didymus*, mit beobachtungen uber einige andere arten der gattung, *Helgol. Wiss. Meersunters.*, 25, 384, 1973.
16. Drebes, G., Sexuality, in *The Biology of Diatoms*, Werner, D., Ed., Univ. of Calif. Press, Berkeley, 1977, chap. 9.
17. Lebour, M. V., *The planktonic diatoms of the northern seas*, Ray Society, London, 1930.
18. Fryxell, G. A., Villareal, T. A., and Hoban, M. A., *Thalassiosira scotia*, sp. nov.: observations on a phytoplankton increase in early austral spring north of the Scotia Ridge, *J. Plankton Res.*, 1, 355, 1979.
19. Drebes, G., The life history of the centric diatom *Bacteriastrum hyalinum* Lauder, *Nova Hedwigia Beih.*, 39, 95, 1972.
20. Fryxell, G. A., Doucette, G. J., and Hubbard, G. F., The genus *Thalassiosira*: the bipolar diatom *T. antarctica* Comber, *Bot. Mar.*, 24, 321, 1981.
21. Hargraves, P. E., Studies on marine planktonic diatoms. I. *Chaetoceros diadema* (Ehr.) Gran: life cycle, structural morphology, and regional distribution, *Phycologia*, 11, 247, 1972.
22. Hargraves, P. E., Studies on marine planktonic diatoms. II. resting spore morphology, *J. Phycol.*, 12, 118, 1976.
23. Hargraves, P. E., Studies on marine planktonic diatoms. IV. morphology of *Chaetoceros* resting spores, *Nova Hedwigia Beih.*, 64, 99, 1979.

24. Hasle, G. R., Some marine planktonic genera of the family Thalassiosiracea, *Nova Hedwigia Beih.*, 45, 1, 1974.
25. Hoban, M. A., Fryxell, G. A., and Buck, K. R., Biddulphoid diatoms: resting spores in Antarctic *Eucampia* and *Odontella, J. Phycol.*, 16, 591, 1980.
26. Syversten, E. E., Resting spore formation in clonal cultures of *Thalassiosira antarctica* Comber, T. *nordenskioeldii* Cleve, and *Detonula confervacea* (Cleve) Gran, *Nova Hedwigia Beih.*, 64, 41, 1979.
27. Stosch, H. A. von and Drebes, G., Entwicklungsgeschichtliche untersuchungen an zentrichen diatomeen. IV. Die planktondiatomee *Stephanopyxis turris*-ihre behandlung und entwicklungsgeschichte, *Helgol. Wiss. Meersunters.*, 11, 209, 1964.
28. Heimdal, B. R., Vegetative cells and resting spores of *Thallassiosira constricta* Gaarder (Bacillariophyceae). *Norw. J. Bot.*, 18, 153, 1971.
29. Heimdal, B. R., Further observations on the resting spores of *Thalassiosira constricta* (Bacillariophyceae). *Norw. J. Bot.*, 21, 303, 1974.
30. Stosch, H. A. von and Fecher, K., "Internal thecae" of *Eunotia soleirolii* (Bacillariophyceae): Development, structure, and function as resting spores, *J. Phycol.*, 15, 233, 1979.
31. Lund, J. W. G., The seasonal cycle of the planktonic diatom, *Melosira italica* (Ehr.) Kutz. subsp. *subarctica* O. Mull., *J. Ecol.*, 42, 151, 1954.
32. Anderson, O. R., The ultrastructure and cytochemistry of resting cell formation in *Amphora coffaeformis* (Bacillariophyceae), *J. Phycol.*, 11, 272, 1975.
33. Hargraves, P. E. and French, F. W., Observations on the survival of diatom resting spores, *Nova Hedwigia Beih.*, 53, 229, 1975.
34. Hargraves, P. E. and French, F. W., Diatom resting spores: significance and strategies, in *Survival Strategies in the Algae*, Fryxell, G. A., Ed., Cambridge University Press, 49, 1983.
35. Fryxell, G. A. and Medlin, L. K., Chain forming diatoms: evidence for parallel evolution in *Chaetoceros. Cryptogamie: Algologie*, 2, 3, 1981.
36. Schmid, A. M., Influence of environmental factors on the development of the valve in diatoms, *Protoplasma*, 99, 95, 1979.
37. Hasle, G. R. and Syvertsen, E. K., The diatom genus *Ceratulina:* morphology and taxonomy, *Bacillaria*, 3, 79, 1980.
38. Pavillard, J., Sur la reproduction du *Chaetoceros eibenii* Meunier, *C. R. Hebd. Séances Acad. Sci., Paris*, 172, 469, 1921.
39. Gran, H. H., The plankton production in the North European waters in the spring of 1912, *Bull. Planktonique pour l'année 1912, Cons. Perm. Explor. Mer.*, 1915.
40. Davis, C. O., Hollibaugh, J. T., Seibert, D. L. R., Thomas, W. H., and Harrison, P. J., Formation of resting spores by *Leptocylindrus danicus* (Bacillariophyceae) in a controlled experimental ecosystem, *J. Phycol.*, 16, 296, 1980.
41. French, F., Diatom resting spores: a comparison of occurrence in the life cycles of *Chaetoceros diadema* (Ehr.) Gran and *Leptocylindrus danicus* Cl., *J. Phycol.*, 16(Suppl. Abs.), 11, 1980.
42. Braarud, T., Microspores in diatoms, *Nature, London*, 143, 899, 1939.
43. Braarud, T., Experimental studies on marine planktonic diatoms, *Avh. Utgitt Norske Vidensk. Akad. Oslo, Mat. Naturvidensk. Kl.*, 10, 1, 1945.
44. Dodson, A. N. and Thomas, W. H., Marine phytoplankton growth and survival under simulated upwelling and oligotrophic conditions, *J. Exp. Mar. Biol. Ecol.*, 26, 153, 1977.
45. Smayda, T. J. and Mitchell-Innes, B., Dark survival of autotrophic, planktonic marine diatoms, *Mar. Biol.*, 25, 195, 1974.
46. Umebayashi, O., Preservation of some cultured diatoms, *Bull. Tokai Reg. Fish. Res. Lab.*, 68, 55, 1972.
47. Matsue, Y., On the culture of the marine planktonic diatom *Skeletonema costatum* (Grev.) Cleve, in Suisangku no Giakan, Jap. Soc. Adv. Sci., 1, 1954; data cited in Smayda, T. J. and Mitchell-Innes, B., *Mar. Biol.*, 25, 195, 1974.
48. Takano, H., Diatom culture in artificial sea water. I. *Bull. Tokai Reg. Fish. Res. Lab.*, 37, 17, 1963.
49. Antia, N. J., unpublished observations, cited by Smayda, T. J. and Mitchell-Innes, B., *Mar. Biol.*, 25, 195, 1977.
50. Antia, N. J. and Cheng, J. Y., The survival of axenic cultures of marine planktonic algae from prolonged exposure to darkness at 20 C, *Phycologia*, 9, 179, 1970.
51. Bunt, J. S. and Lee, C. C., Data on the composition and dark survival of four sea-ice microalgae, *Limnol. Oceanogr.*, 17, 458, 1972.
52. Rodhe, W., Can phytoplankton production proceed during winter darkness in subarctic lakes?, *Int. Ver. Theor. Angew. Limnol. Vehr.*, 12, 117, 1955.
53. Hellebust, J. A. and Lewin, J., Heterotrophic nutrition, in *The Biology of Diatoms*, Werner, D., Ed., Univ. of Calif. Press, Berkeley, 1977, 169.

54. Ackley, S. F., Taguchi, S., and Buck, K. R., Primary productivity in sea ice of the Weddell Sea, CRREL Report 78-19, U.S. Army Corps. of Engineers, Hanover, New Hampshire, 1978.
55. Schandelmeier, L. and Alexander, V., An analysis of the influence of ice on spring phytoplankton population structure in the southeast Bering Sea, *Limnol. Oceanogr.*, 26, 935, 1981.
56. Garrison, D. L. and Buck, K. R., unpublished data, 1982.
57. Fryxell, G. A., personal communication, 1982.
58. Wood, E. J. F., Diatoms in the ocean depths, *Pac. Sci.*, 10, 377, 1956.
59. Kimball, J. F., Jr., Corcran, E. F., and Wood, E. J. F., Chlorophyll-containing microorganisms in the aphotic zone of the ocean, *Bull. Mar. Sci. Gulf Caribb.*, 13, 574, 1963.
60. Venrick, E. L., The distribution and ecology of oceanic diatoms in the North Pacific, Ph.D. thesis, Univ. of Calif., San Diego, 1969.
61. Smayda, T. J., Normal and accelerated sinking rates of phytoplankton in the sea, *Mar. Geol.*, 11, 105, 1971.
62. Venrick, E. L., McGowan, J. A., and Mantyla, A. W., Deep maxima of photosynthetic chlorophyll in the Pacific Ocean, *Fish. Bull., U.S.*, 71, 41, 1973.
63. Malone, T. C., Garside, C., Anderson, R., Roels, O. A., The possible occurrence of photosynthetic microorganisms in deep-sea sediments of the north Atlantic, *J. Phycol.*, 9, 482, 1973.
64. Silver, M. W. and Alldredge, A. L., Bathypelagic marine snow: deep-sea algae and detrital community, *J. Mar. Res.*, 39, 501, 1981.
65. Smayda, T. J., Biogeographical studies of marine phytoplankton, *Oikos*, 9, 158, 1958.
66. Wood, E. J. F., Some aspects of marine microbiology, *J. Mar. Biol. Assoc. India*, 1, 26, 1959.
67. Roman, M. R. and Tenore, K. R., Tidal resuspension in Buzzards Bay, Massachusetts. 1. seasonal change in the resuspension of algal carbon and chlorophyll a, *Estuarine Coastal Mar. Sci.*, 6, 37, 1978.
68. Walsh, J. J., Whitledge, T. E., Barvaenik, F. W., Wirick, C. D., Howe, S. O., and Esaias, W. E., Wind events and food chain dynamics within the New York Bight, *Limnol. Oceanogr.*, 23, 659, 1978.
69. Malone, T. C. and Chervin, M. B., The production and fate of phytoplankton size fractions in the plume of the Hudson River, New York Bight, *Limnol. Oceanogr.*, 24, 683, 1979.
70. Takahashi, M., Seibert, D. L., and Thomas, W. H., Occasional blooms of phytoplankton during summer in Saanich Inlet, B. C., Canada, *Deep-sea Res.*, 24, 775, 1977.
71. Stosch, H. A. von and Kowallik, K., Der von L. Geitler aufgestellte saltz uber die notwendigkeit einer mitose fur jebe schalenbildung von diatomeen. Beobachtungen uber die reichweite und uberlegungen zu seiner zellmechanischen bedeutung, *Ost. Bot. Z.*, 116, 454, 1969.
72. Durbin, E. G., Aspects of the biology of resting spores of *Thalassiosira nordenskioeldii* and *Detonula confervacea*, *Mar. Biol.*, 45, 31, 1978.
73. Hollibaugh, J. T., Seibert, D. L., and Thomas, W. H., Observations on the survival and germination of resting spores of three *Chaetoceros* (Bacillariophyceae) species, *J. Phycol.*, 17, 1, 1981.
74. French, F. W. and Hargraves, P. E., Physiological characteristics of planktonic diatom resting spores, *Mar. Biol. Lett.*, 1, 185, 1980.
75. Doucette, G., The bipolar diatom *Thalassiosira antarctica* Comber and *Porosira glacialis* (Grunow)Jørgensen; comparative ultrastructure of vegetative and resting stages of disjunct populations, M.S. Thesis, Texas A&M University, 1982.
76. Anderson, R. O., Respiration and photosynthesis during resting cell formation in *Amphora coffaeformis* (Ag.) Kutz, *Limnol. Oceanogr.*, 21, 452, 1976.
77. Gran, H. H., Pelagic plant life, in *The Depths of the Ocean*, Murray, J. and Hjort, J., Eds., Macmillan, London, 1912, chap. 6.
78. Gross, F., The life history of some marine planktonic diatoms, *Philos. Trans. R. Soc.* London B, 228, 1, 1937.
79. Hardy, A. C. and Gunther, E. R., The plankton of the South Georgia whaling grounds and adjacent waters 1926—7, *Discovery Rep.*, 11, 1, 1935.
80. Hart, T. J., On the phytoplankton of the south-west Atlantic and Bellinghausen Sea, *Discovery Rep.*, 8, 1, 1934.
81. Hart, T. J., Phytoplankton periodicity in Antarctic surface waters, *Discovery Rep.*, 21, 261, 1942.
82. Garrison, D. L., Studies of coastal phytoplankton populations in Monterey Bay, California, Ph.D. dissertation, Univ. of Calif., Santa Cruz, 1980.
83. Garrison, D. L., Monterey Bay phytoplankton. II. Resting spore cycles in coastal diatom populations, *J. Plankton Res.*, 3, 137, 1981.
84. Ikari, J., On the formation of auxospores and resting spores of *Chaetoceros teres* Cleve, *Bot. Mag., Tokyo*, 35, 222, 1921.
85. Hasle, G. R. and Smayda, T. J., The annual phytoplankton cycle at Drobak, Oslofjord, *Norw. J. Bot.*, 8, 53, 1960.

86. Smayda, T., A quantitative analysis of the phytoplankton of the Gulf Panama. III. General ecological conditions and the phytoplankton dynamics at 8°45′ N, 79°23′ W from November 1954 to May 1957, *Bull. Inter-Amer. Trop. Tuna Comm.*, 11, 355, 1966.

87. Sakshauge, E. and Mykelstad, S., Studies on the plankton ecology of the Trondheimfjord. III. Dynamics of the phytoplankton blooms in relation to environmental factors, bioassay experiments, and parameters for the physiological state of the population, *J. Exp. Mar. Biol. Ecol.*, 11, 157, 1973.

88. Mikhailova, N. F., The germination of resting spores of *Chaetoceros lauderi* Ralfs, *Dokl. Akad. Nauk SSSR*, 143, 741, 1962.

89. Hasle, G. R., An analysis of the phytoplankton of the Pacific Southern Ocean: abundance, composition, and distribution during the Brategg Expedition, 1947—1948, *Havalrad Skr.*, 52, 6, 1969.

90. Beinfang, P. K., Sinking rates of heterogeneous, temperate phytoplankton populations, *J. Plankton Res.*, 3, 235, 1981.

91. Zgurovskaya, L. N., The effect of the addition of nutrients on the growth of spores and the division of planktonic algae from bottom sediment, *Oceanology*, 17, 119, 1977.

92. Huyer, A. and Smith, R. L., Physical characteristics of Pacific Northwest coastal waters, in *The Marine Plant Biomass of the Pacific Northwest Coast*, Krauss, R. W., Ed., Oregon State Univ. Press, 1977, chap. 3.

93. Bryden, H. L., Mean upwelling velocities on the Oregon continental shelf during summer 1973, *Estuarine Coastal Mar. Sci.*, 7, 311, 1978.

94. Silver, M. W., Shanks, A. L., and Trent, J. D., Marine snow: microplanktonic habitat and source of small-scale patchiness in pelagic populations, *Science*, 201, 371, 1978.

95. Alldredge, A. L., The chemical composition of microscopic aggregates in two neritic seas, *Limnol. Oceanogr.*, 24, 855, 1979.

96. Shanks, A. L. and Trent, J. D., Marine snow: sinking rates and potential role in vertical flux, *Deep-sea Res.*, 27, 137, 1980.

97. Kashkin, N. I., On the winter deposits of phytoplankton in the sublittoral, *Tr. Inst. Okeanol. Inst. Akad. Nauk SSSR*, 65, 49, 1964.

98. Davis, J. S., Survival records in the algae, and the survival role of certain algal pigments, fats, and mucilaginous substances, *The Biologist*, 54, 52, 1972.

99. Riley, G. A., Physiological aspects of spring diatom flowering, *Bull. Bingham Oceanogr. Coll.*, 8, 1, 1943.

100. Sussman, A. S., Physiology of dormancy and germination in the propagules of cryptogamic plants, *Handb. Pflphysiol.*, 15, 933, 1965.

101. Garrison, D. L., Monterey Bay phytoplankton I. Seasonal cycles of phytoplankton assemblages, *J. Plankton Res.*, 1, 241, 1979.

102. Tont, S. A., Short-period climatic fluctuations: effects on diatom biomass, *Science*, 194, 942, 1976.

103. French, F. W. III. and Hargraves, P. E., Interclonal differences in ability to form resting spores, *Limnol. Oceanogr.*, (Abstracts of papers for forty-second annual meeting, June 1979).

Chapter 2

LIFE HISTORIES, DISPERSAL, AND SURVIVAL IN MARINE, PLANKTONIC DINOFLAGELLATES

Linda M. Walker

TABLE OF CONTENTS

I. INTRODUCTION

Dinoflagellates are a geologically old group with a fossil record dating back to the Upper Silurian. Today this diverse group is represented by 1000 to 1500 species in marine and freshwater habitats worldwide. The majority are free-living planktonic and benthic species, but symbiotic and parasitic species also occur (Figure 1). Most free-living forms are photosynthetic and therefore are found in the euphotic zone. Heterotrophs and resting stages usually occur below this zone or in shallow water sediments with lowered light levels. Photosynthetic dinoflagellates contribute to the base of the food web as primary producers; a few are mixotrophic and capable of phagocytosis. Some dinoflagellates, such as *Kofoidinium* and *Polykrikos*, prey on other dinoflagellates, copepod eggs, and nauplii and fish eggs. Dinoflagellates are preyed upon by a variety of invertebrates and larval and adult fish.

Despite the diversity of habitats and modes of existence, all dinoflagellates share a number of unique characters. Dinoflagellates are single-celled although some may form chains or pseudocolonies. The size of individual cells ranges from approximately 5µm to >2mm. All dinoflagellates have a unique nucleus at some stage in their life history. This nucleus contains continually condensed chromosomes and is termed dinocaryotic. Another characteristic of all dinoflagellates, except *Noctiluca*, is the occurrence, at some stage, of two dissimilar flagella. The longitudinal flagellum and the transverse flagellum, which lies in the groove or cingulum surrounding the cell, work together to propel and stabilize the cell. Photosynthetic species usually have a distinct xanthophyll pigment, peridinin. Dinoflagellates can be categorized into two broad groups based on cell wall structure. Armored forms have cellulose or other polysaccharide plates in vesicles between membranes. Unarmored forms lack these plates. This separation can be made fairly readily at the light microscope level but, ultrastructurally, gradations in the number and thickness of plates in different species obscure the distinction. Nevertheless, this separation is useful when discussing life histories because of the morphological implications of the presence or absence of these plates.

All dinoflagellates reproduce asexually by mitotic division and an increasing number are known to reproduce sexually as well. Nevertheless, sexual reproduction may not occur in all genera, particularly the oceanic forms. All dinoflagellates, except one, have a life history in which all stages except the zygote, where it occurs, have a haploid number of chromosomes. In *Noctiluca scintillans* the life history is diplontic where all stages are diploid except during gametogenesis when meiosis produces haploid gametes.

Noctiluca miliaris is currently an anomaly among the dinoflagellates. Observations in the literature continually challenge its designation as a dinoflagellate. Omori and Hamner[11] observed an unusual, periodic feeding pattern in a tropical lagoon involving the aggregation of thousands of *N. miliaris* and the production of sinking mucoid webs to trap food particles. In addition to its diplontic life history, *N. miliaris* has a unique nucleus in the vegetative stage which is unlike a dinocaryotic nucleus. It also lacks a transverse flagellum in every stage of its life history. During gametogenesis, synchronous division proceeds after the completion of the two meiotic divisions to produce up to 1024 haploid gametes within the mother cells.[12] The gametes do not swim with characteristic rotation of the body seen in other dinoflagellates. The resemblance of the gametes to a *Gymnodinium* cell is only superficial. Although the gametes possess a nucleus that is considered dinocaryotic, the dinoflagellate affinities of *Noctiluca* seem tenuous and it will be excluded from further discussion in this chapter.

The importance of life histories in understanding and describing dinoflagellate survival and dispersal has been demonstrated by recent work with dinoflagellates that form blooms.[13-16] Studies of several bloom species have demonstrated relationships

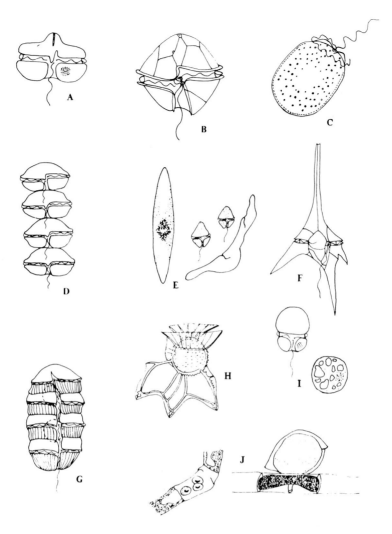

FIGURE 1. Representative dinoflagellates. (Not drawn to scale.) (A) *Ptychodiscus brevis*, an unarmored toxic bloom species (after Walker[1]). (B) *Protogonyaulax tamarensis*, an armored toxic bloom species (after Loeblich and Loeblich[2]). (C) *Prorocentrum lima*, a view of the left valve (after Steidinger[3]). (D) *Gonyaulax monilata*, a toxic chain-forming species (after Steidinger and Williams[4]). (E) *Pyrocystis acuta*. Two stages of the asexual life cycle of this oceanic species — the coccoid form and the thecate, biflagellated planospores (after Elbrächter and Drebes[5]). (F) *Ceratium hirundinella*, a freshwater species (after Wall and Evitt[6]). (G) *Polykrikos kofoidii*, a pseudocolonial species (after Kofoid and Swezy[7]). (H) *Ornithocercus magnificus*, an oceanic species with elaborate flanges and lists (after Taylor[8]). (I) *Symbiodinium microadriatica*, the flagellated zoospore and vegetative cell of this invertebrate symbiont (after McLaughlin and Zahl[9]). (J) *Cystodinium bataviense*, a freshwater parasite. Amoebae within an *Oedogonium* cell and a vegetative cell penetrating an *Oedogonium* cell (after Pfiester and Lynch[10]).

between life history strategies and bloom initiation, bloom termination, transport, dispersal, and survival during winter or other forced resting periods. Similar relationships between life history strategies and dispersal and survival probably hold true for a number of dinoflagellates.

Most life history work has concentrated on neritic and freshwater forms. Life history information for oceanic, benthic, symbiotic, and parasitic forms is scarce because of difficulties in obtaining specimens and providing specialized culture conditions. This chapter will deal mainly with neritic, free-living species, although references will be made to other dinoflagellate groups for clarification and comparison.

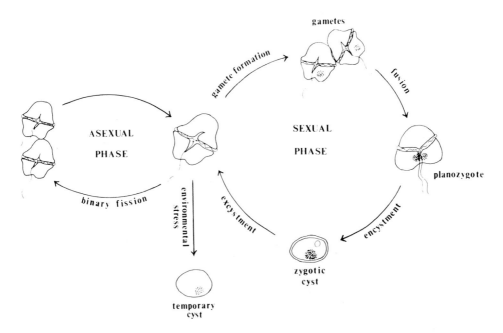

FIGURE 2.   Representative life cycle of a free-living dinoflagellate. The sexual phase has been documented for 23 species. Life histories are more complex in parasitic, symbiotic and oceanic species.

II. ASEXUAL LIFE HISTORY

In simplest terms, dinoflagellates reproduce asexually by binary fission. Fission is transverse or oblique and, in dividing pairs, the cingula are usually parallel to each other (Figure 2). The diversity of dinoflagellates is reflected in the various modes of asexual reproduction in this group. The differences can be accounted for by differences in mode of existence (free-living vs. parasitic vs. symbiotic), morphology (armored vs. unarmored; chain-formers vs. solitary cells), and habitat.

In solitary, unarmored dinoflagellates, division typically produces two identical daughter cells of smaller initial size. In the genus *Polykrikos*, cytokinesis is incomplete and pseudocolonies of 2 to 16 subunits result; karyokinesis does not always accompany cytokinesis.

Several basic patterns are found among armored dinoflagellates. One pattern of division divides the theca between the daughter cells and each adds the missing plates during growth. This division pattern is found in *Ceratium*[4] and *Pyrodinium bahamense*[17] among others. In other genera, such as some freshwater and marine *Peridinium*[18-21] and *Glenodinium*[22] the parent theca splits and releases a naked cell which divides into two or more daughter cells. These daughter cells then produce new thecae. In some species of the oceanic genus *Pyrocystis* the armored, biflagellated daughter cells (planospores) form and divide within a nonmotile, coccoid cell.[5,23] The daughter cells are released and then quickly shed their thecae and swell to the size of the parent cell. In other *Pyrocystis* species the daughter cells may be unarmored, biflagellated or uniflagellated planospores or nonmotile, unarmored aplanospores. In some chain-forming species, such as *Gonyaulax monilata*, the daughter cells in nature do not normally separate unless physically disturbed, and can form chains up to 32 cells or more. In culture, this species can produce larger, solitary cells. The freshwater, parasitic *Cystodinium bataviense* reproduces asexually by aplanospores, zoospores and amoebae.[10]

Division is initiated at certain times of the day, and the actual time varies with species. Marine, photosynthetic dinoflagellates generally divide in late night or early morning.[24] Buchanan[17] found that in Oyster Bay, Jamaica, *Pyrodinium bahamense* var. *bahamense* initiated division after sunrise, with the largest number of dividing pairs found one to two hours after sunrise. Weiler and Eppley[24] reported that marine *Ceratium* spp. under similar laboratory and field conditions had comparable initiation times, shortly before or after dawn. Coats and Heinbokel[25] used acridine orange to stain nuclei and counted dividing pairs in *Ceratium furca.* They found that the highest numbers of dividing pairs occurred shortly after dawn in Chesapeake Bay. Under laboratory conditions, 85% of all division in *Gonyaulax polyedra* took place in a five-hour period spanning the end of the dark and the beginning of the light period.[26]

Hastings and Sweeney[27] have termed this synchronization of binary fission "phased cell division". Cells which are physiologically able to divide do so during this time period and those which are not, wait for the next cycle. Since division and hence population growth occur only during certain time periods, the growth curve is stepwise. A calculated curve for the stepwise growth curve gives a measure of the average generation time or the average growth rate. The phased division is a function of endogenous, and to some extent exogenous, controls. The biological clock which keys rhythmic cell division to a 24-hour cycle under natural conditions can be manipulated in laboratory cultures. In laboratory cultures of *G. polyedra* various light levels can be used to shorten, dampen or erase the periodicity.[26,27] Temperature does not affect the period of cell division rhythm although it can affect the average rate of cell division.

Growth rates vary widely with species and environmental conditions. Weiler and Eppley[24] observed a maximum growth rate for laboratory cultures of *Ceratium furca* of one doubling per day. Weiler[28] studied 51 species of *Ceratium* in the oligotrophic waters of the North Pacific central gyre and found rates of 5.2 to 7.5 days for doubling. Swift et al.[29] reported 15 days for *Pyrocystis noctiluca* and 4 to 5 days for *P. fusiformis* to double in natural populations off the West Indies. Seliger et al.[30] observed 3 to 3.5 days for doubling for *Pyrodinium bahamense* var. *bahamense* in natural populations in Oyster Bay, Jamaica. Tyler and Seliger[31] found growth rates of .05 to .8/day for *Prorocentrum minimum* (var. *mariae-lebouriae*) depending on environmental conditions of temperature and salinity. Rates of .06 to .3/day for *Protogonyaulax tamarensis* have been reported under natural conditions.[32]

Division occurs in cells which are at optimum physiological states under suitable physical conditions of light, temperature, nutrients, salinity, etc. If environmental conditions become unsuitable, dinoflagellates may form a temporary resting stage to survive. A shock due to rapid change in conditions, such as a sudden rise or fall in temperature, light, or salinity, can cause the cell to lose its flagella and become immobile. Unarmored forms frequently swell. The cell can maintain this state for a limited time and, if suitable conditions are restored, will often recover.[33] A slower deterioration of physical conditions can lead to the formation of a temporary cyst. The flagella are shed and, in some cases, mucilage is produced. Such cysts may aid survival until suitable conditions return. Temporary cysts are not long term solutions to environmental changes but they may allow organisms to survive high temperatures in summer and short-term salinity changes. Some freshwater species use asexual cysts as overwintering stages.[34,35]

III. SEXUAL LIFE HISTORY

An increasing number of dinoflagellates are known to have sexual life cycles[1,10,13,14,16,19-21,36-49] including symbiotic and parasitic forms (Table 1). The number will no doubt increase as other species are studied utilizing the increased knowledge of

Table 1
DINOFLAGELLATES WITH KNOWN SEXUAL CYCLES

Species	Habitat	Mode of existence	Field or culture observation	Ref.
Amphidinium carterae	marine	free-living	culture	36
Ceratium cornutum	freshwater	free-living	culture	37
C. horridium	marine	free-living	culture	37
Crypthecodinium cohnii	marine	free-living	culture	38
Cystodinium bataviense	freshwater	parasitic	culture	10
Gambierdiscus toxicus	marine	free-living	culture	39
Glenodinium lubiniensiforme	freshwater	free-living	culture?	40
Gonyaulax monilata	marine	free-living	culture/field	41
Gymnodinium fungiforme	marine	free-living	culture	45
G. pseudopalustre	freshwater	free-living	culture/field	46
Gyrodinium uncatenum	marine	free-living	field	16
Helgolandidinium subglobossum	marine	free-living	culture	37
Oxyrrhis marina	marine	free-living	culture	37
Peridinium cinctum f. ovoplanum	freshwater	free-living	culture	19
P. gatunense	freshwater	free-living	culture	21
P. willei	freshwater	free-living	culture	20
P. volzii	freshwater	free-living	culture	47
Protogonyaulax catenella	marine	free-living	culture	48
P. tamarensis	marine	free-living	culture/field	13,14,42-44
Ptychodiscus brevis	marine	free-living	culture/field	1
Symbiodinium microadriaticum	marine	symbiotic	culture	49
Woloszynskia apiculata	freshwater	free-living	culture	46

the sexual cycle and techniques available to induce sexual reproduction in culture populations.[1,19-21,41,46] One would expect meiosis to be important to long-term species survival in organisms with complexly packed chromosomes, despite the polytenic nature of the chromosomes. Bernstein asserts that in a variety of prokaryotic and eukaryotic organisms sexual reproduction functions primarily to repair damaged DNA through recombinational repair.[50] However, Steidinger[3] and Dale[51] point out that some dinoflagellates may not reproduce sexually. Oceanic species, in particular, may not have a sexual phase or may have an abbreviated sexual phase which lacks a hypnozygote. Steidinger[3] observes that it would be counterproductive for an oceanic species to produce a benthic hypnozygote which had little chance of returning to the motile population.

The majority of work and observation on sexual phases of dinoflagellate life histories has come from laboratory cultures. Work with field populations has provided confirmatory results in some cases. The use of laboratory cultures is almost a prerequisite since production of sexual stages may cover several weeks. Depending on field sampling to provide continuous data is unreasonable in most cases considering the motility of dinoflagellates and physical conditions such as wind and currents unless a discrete small water mass could be continually monitored. Sexual stages can occur spontaneously in dense or old laboratory cultures or can be induced by various techniques. Laboratory conditions can only, at best, approximate field conditions necessary to induce gamete production, encystment and excystment. The low frequency of hypnozygote development or lack of hypnozygotes in some dinoflagellate cultures may be due to this fact or to clonal cultures of one mating type.

The generalized sexual cycle is illustrated in Figure 2. Under appropriate conditions, gametes are produced which fuse to form a diploid planozygote. The planozygote can be readily identified under the light microscope by two longitudinal flagella. The planozygote may undergo meiosis or, in most cases, develop into a resting, resistant cyst or hypnozygote. The hypnozygote is characterized by a thick, resistant cell wall and, in most species, requires a dormancy period. Meiosis can occur within the cyst or after excystment to restore the vegetative haploid condition.

Gametes have been observed in dense, actively growing or old, presumably nutrient depleted laboratory cultures.[1,19-21,41] Gametes have also been observed in field blooms of *Protogonyaulax tamarensis*,[14] *Gyrodinium uncatenum*,[16] and *Ptychodiscus brevis*.[1] Manipulation of laboratory cultures has also been used to induce gamete formation. Von Stosch[46] used a shortened light period to induce gamete production in actively growing cultures of the freshwater *Gymnodinium pseudopalustre*. Pfiester[19-21] and Walker and Steidinger[41] utilized nitrogen deficient media to induce gametes in several freshwater and one marine dinoflagellate. Blue light[1] and mixing clonal cultures of heterothallic species[1,19-21,46-48] also have been used to induce gamete formation. Induction usually occurs within several days but can occur in as little as 15 minutes.[47] The mechanism(s) by which these stimuli effect sexual induction is at present unknown. Furthermore, these stimuli may have no relation to natural conditions. Studies with the freshwater volvocid *Chlamydomonas* have shown that reduced nitrogen levels can induce gametogenesis in some, but not all species in laboratory cultures. However under natural conditions, sexual reproduction occurs under even high levels of nitrogen.[52,53]

As in the asexual phase, the sexual phase differs in details among dinoflagellate species. Gametes are generally recognizably smaller than the vegetative cell. Von Stosch[54] attributes this to "depauperate mitosis" during which cells divide before cytoplasmic synthesis is completed, resulting in smaller division products. Gametes may be the same size, termed isogamous, or of different size, termed anisogamous. Gametes may be darker or lighter than the vegetative cell. The nucleus is very prominent. Gametes of armored forms may or may not be armored. Gametes can represent mating strains.

Fusion usually occurs during the dark cycle in laboratory cultures. Just prior to fusion, "dancing" behavior may occur during which clusters of single gametes aggregate, then pair off.[1,46] The actual site of contact between gametes seems species specific and may involve a "copulation globule" as noted for several freshwater species by Von Stosch.[46]

Fusing pairs can generally be distinguished from dividing pairs since the cingula are perpendicular during fusion, although in some species the cingula are parallel. Fusion can be rapid. Fusion in the parasitic *Cystodinium bataviense* occurs in four to five minutes.[10] In contrast, Turpin et al.,[44] reported that fusion in *Protogonyaulax tamarensis* required several weeks under laboratory conditions. Generally, fusion requires several hours. In some species, such as *Cystodinium*,[10] the armored gametes shed their thecae during fusion.

The resulting planozygote is characterized by two longitudinal flagella throughout its development. One of the transverse flagella is thought to be lost or resorbed at some point.[46] During this stage, the nuclei migrate to lie beside each other and begin the fusion process. The planozygote in most species enlarges and darkens in color. *Amphidinium carterae* is an exception.[36] The planozygote may settle immediately, as in *A. carterae,* or remain motile for a varying amount of time. Eventually the planozygote sheds its flagella and develops a thickened, three-layered cell wall. This encysted stage is termed the hypnozygote. In some dinoflagellates, such as *Ceratium horridum*, this stage may be skipped and the planozygote undergoes meiosis.[37] Whether this is due to culture conditions or is natural is unknown.

Hypnozygote walls can be smooth or have processes of various shape, length, and number. Some species also have a mucilage layer surrounding the cyst wall. Species identification of cysts is difficult but involves shape, size, wall morphology, and cytoplasmic characteristics.

The thick cyst wall confers extraordinary resistance to the zygote. The wall is resistant to bacterial degradation, temperature and salinity extremes, anoxic conditions and to some degree, dessication. The wall is an effective barrier against penetration of many stains and fixatives; this has hindered, to some extent, the study of cyst development.

A pigment body (often red or gold) is often present and reduces in size as the hypnozygote ages. The cytoplasm generally pulls away from the wall during maturation and in some species may continue to darken.

Most species seem to have a period of required dormancy before excystment can occur. Any attempts to induce excystment during this period is usually unsuccessful. The time required for dormancy varies between species. *Peridinium gatunense* required 12 hr;[21] *P. cinctum* and *P. willei* required 7 to 8 weeks;[19,20] *Gonyaulax monilata* cysts required 3 to 4 weeks under laboratory conditions.[41] *Protogonyaulax tamarensis* cysts excysted after 1 to 4 months.[14,44] Anderson's[14] experiments point to the importance of temperature during dormancy; cooler temperatures produce longer dormancy periods. This probably relates to lowered metabolism rates and the necessary biochemical and physiological changes which must occur within the cyst before excystment can occur. *Protogonyaulax tamarensis* cysts stored at 22°C used starch reserves more rapidly than those stored at 5°C, indicating a higher metabolic rate at 22°C.

Triggers for excystment after the dormancy period may include a change in temperature, light, some endogenous trigger such as food reserves or a combination of endogenous and exogenous triggers. Anderson[14] found that *P. tamarensis* cysts stored at 5°C and 22°C responded to an increase and decrease, respectively, in temperature. The presence or absence of light had no effect.

Work on freshwater dinoflagellates indicates the timing of meiotic division in the sexual phase is variable. In *Peridinium cinctum*, *P. willei* and *P. gatunense*, meiosis occurs prior to germination. Three nuclei abort in *P. cinctum* and *P. willei* and one haploid cell germinates; in *P. gatunense* only two nuclei abort and two haploid cells germinate.[19-21] Meiosis in *Gymnodinium pseudopalustre* occurs after germination.[46] One biflagellated cell is released and divides to form four daughter cells. In *Woloszynskia apiculata* the first meiotic division occurs within the cyst releasing two daughter cells which then undergo the second meiotic division. Sometimes this second division occurs prior to germination and four cells are released.[46] Meiosis in *Ceratium cornutum* is also after germination.[37]

IV. DISPERSAL AND SURVIVAL

Understanding the dispersal and survival of dinoflagellates requires knowledge of the organisms' behavior, life cycle strategies, physiology, the physical factors of the organisms' habitat and the relationships between all these factors. Dispersal, or the lack of dispersal, and survival of populations and therefore species, are to a large extent interrelated and will be considered together in this section.

Dispersal mechanisms include wind, tides, currents, density fronts and upwelling. Human agents of dispersal may include transportation by ship ballast, bilge water, on ship hulls, and on transplanted shellfish. Dispersal can involve any life history stage, motile or resting. Although dinoflagellates are motile, long range dispersal depends on physical or a combination of physical and biological mechanisms. Biological mecha-

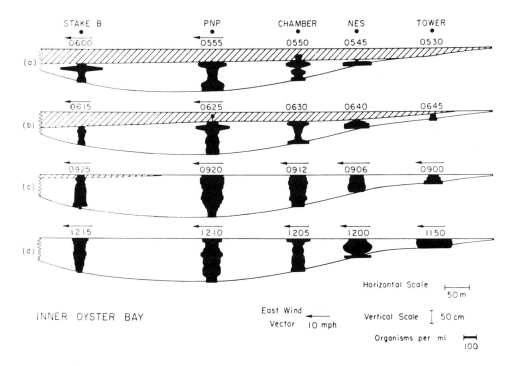

FIGURE 3. Salinity, wind, and depth distributions of *Pyrodinium bahamense* var. *bahamense* concentrations in Oyster Bay, Jamaica, along an east-west transect during the morning of 26 July 1968. The hatched water layer represents the brackish layer (salinity=10 to 20 ′/″); the clear water layer represents the saline layer (salinity ≥ 30 ′/″). The westward pointing arrows are wind velocity vectors. The black areas indicate the concentration of *P. bahamense*; the width at any depth is proportional to cell concentration. Throughout the morning the westward wind, the displacement of the brackish layer, and the phototactic response of *P. bahamense* combine to concentrate the dinoflagellates in the eastern-most part of the bay. (From Seliger et al., *Limnol. Oceanogr.*, 15, 234, 1970. With permission.)

nisms can often enhance physical dispersal. Diurnal migration, positive or negative phototaxis, and the timing of life history stages are examples.

Although a number of marine dinoflagellates are distributed worldwide or over a relatively large geographical area, not all dinoflagellates utilize the above mechanisms in the strict sense. In estuarine species these mechanisms usually limit dispersal and retain populations within estuarine habitats.

The daily accumulation mechanisms that retain *Pyrodinium bahamense* var. *bahamense* in high concentrations in Oyster Bay, Jamaica, involve differential movement of water masses of different densities and a temporal sequence of positive phototaxis (Figure 3).[30] Oyster Bay receives freshwater input from the Martha Brae River. In the early morning a vertical gradient exists. There is a surface brackish layer, a middle layer of intermediate salinity, and a more saline bottom layer. When the daily easterly winds begin to blow, the freshwater layer is forced westward out of the bay, and the intermediate salinity layers flow eastward to compensate. The intermediate layers mix with the more saline bottom waters and the bay becomes uniformly mixed. In the late afternoon, the wind dies and a flood tide pushes the brackish water eastward. The uniformly mixed saline waters are pushed westward and more saline outer bay waters move eastward along the bottom to reestablish the vertical gradient.

Imposed upon this daily movement of water masses is the positive phototactic response of *P. bahamense* var. *bahamense*. In the morning before the easterly wind and vertical mixing begins, this species concentrates in the intermediate layer, beneath the

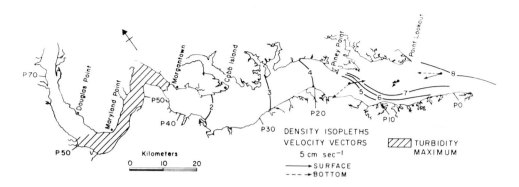

FIGURE 4.   Surface water contours of σ, density isopleths from slack water runs for the Potomac River, November, 1979. Frontal region is delineated by surfacing of 5, 6 and 7, σ, isopleths near the mouth of the river and marks the separation of estuarine circulation and bay water inflow. Turbidity maximum is indicated by the hatched region and marks the mixing of fresh and saline waters. (From Tyler et al., *Mar. Ecol. Prog. Ser.* 7, 163, 1982. With permission.)

brackish layer. *Pyrodinium bahamense* var. *bahamense* will not migrate into the brackish layer. The net effect is that each day a significant portion of the population is deposited in the easternmost portion of the bay, where flushing rates are lowest. The high cell concentrations in this portion of the bay provide brilliant, bioluminescent displays. The mean exchange rates measured for the entire bay would easily flush *P. bahamense* var. *bahamense* from this system without the above accumulating mechanisms.

Pyrodinium bahamense var. *bahamense* provides a relatively simple example of the interaction between physical and behavioral accumulation mechanisms, on a daily basis. Study of bloom species have provided insights into retention and dispersal mechanisms involving life history strategies over an annual cycle.

Bloom species provide a convenient field research tool since the massive numbers involved allow mapping and sampling over a period of time. However, caution must be used in extrapolating mechanism relationships from bloom to non-bloom species. Blooms can often cause catastrophic effects within the affected areas due to toxins and/or oxygen depletion. As a result, blooms are often viewed as perturbations of a stable system and the causative species are considered aberrant. However, another interpretation might be that bloom species are highly adapted at synchronizing biological with physical dispersal/retention mechanisms.

Tyler et al.[16] presented a synthesis of the study of the life history and physiology of *Gyrodinium uncatenum* and the hydrography of the Chesapeake Bay system to explain the annual occurrence of this summer bloom species. The Potomac River is a major Chesapeake Bay tributary and was studied in detail with relation to the *Gyrodinium* blooms. The Potomac River is characterized as a partially mixed coastal estuary (Figure 4). Proceeding downstream toward the mouth region, the river can be divided into regions or zones based on the relative amount of fresh and saline water and the amount of mixing. The position and extent of these zones depend on a number of factors including amount of runoff, tidal amplitude and storm surges. At some point upstream there is a zone of mixing of fresh and saline water. In this transition zone, turbidity is at a maximum. Upstream of this zone, the net flow of freshwater is downstream at all depths. Downstream of this zone is the estuarine circulation zone with net outflow of low salinity surface waters and net up-estuary flow of higher salinity bottom waters. Near the mouth another transition zone occurs which separates the estuarine circulation from bay inflow. In this region, the major river pycnocline surfaces and a frontal convergence marks the downstream extent of estuarine circulation. The pycnocline also tilts with respect to the cross-stream view of the river and intersects the bottom in shoaling areas.

In late summer of 1979 and 1980, Tyler and coworkers observed *Gyrodinium uncatenum* in the estuarine portion of the river with greatest concentrations near the frontal convergence zone. Cells were not found on the bay side of the frontal zone although laboratory experiments indicated cells had comparable growth rates in water from both sides of the front.

This distribution is explained by a combination of behavioral response and hydrographical patterns within the river. Upstream from the convergence zone, *G. uncatenum* exhibits strong positive phototaxis, concentrating in surface waters during midday. Before sunset, *G. uncatenum* migrates downward and remains in the pycnocline until sunrise. This pycnocline boundary probably slows the seaward transport of *G. uncatenum* by surface waters and aids retention within the estuary. Near the convergence zone the strong positive phototaxis and the downwelling flow may result in the accumulation of cells at the frontal interface.

When convergence velocities are greater than the migration speed of *G. uncatenum*, cells may become entrained along the frontal interface and occur below the surface within the convergence. If convergence velocities are lower than migration speed or the front is disrupted, cells may be able to cross the boundary downstream. However, the net upstream flow of surface waters would return the cells to the estuarine zone.

In late summer-autumn, near the end of the bloom season, sexual stages were observed. Early gametes occurred within the convergence zone. Greatest concentrations of late planozygotes/early hypnozygotes were also found within the convergence zone. Across the convergence, concentrations were highest where the pycnocline intersected the bottom in shoaling areas. Upstream from the front, high concentrations were also observed along the bottom, dropping out of the pycnocline. The gametic and zygotic stages did not exhibit diurnal migration and their distribution was associated with a particular water density. As cyst formation continued, motility was lost and cysts settled to the bottom. The deposition of cysts in the sediments reflected the position of the frontal convergence in both years (Figure 5). Highest concentrations of cysts occurred near the mouth of the river along the southern shore where the pycnocline intersected the bottom. Presumably, the cysts overwinter in the sediments and excyst in spring or early summer to resume the bloom cycle.

Another bloom species which has provided numerous insights is *Protogonyaulax tamarensis*. Part of the geographic range of this toxic dinoflagellate includes the estuarine and near coastal waters along the coast of New England and eastern Canada[55-57] where it blooms spring through fall.[58] The toxin can be accumulated by shellfish in bloom areas and contaminated shellfish can cause paralytic shellfish poisoning (PSP) in humans. PSP can be fatal and blooms of this dinoflagellate pose major problems to human health and commercial shellfish industries. Considerable research has concentrated on understanding the population dynamics and life history of the species and associated hydrographic factors.

Central to understanding the bloom phenomenon of *Protogonyaulax tamarensis* is the hypnozygote. Mapping studies along the New England coast have located both estuarine and offshore hypnozygote or cyst accumulations.[57,59] Anderson and Morel[58] proposed a plausible relationship between biological and physical factors to explain blooms in certain tidal estuaries in the Cape Cod, Massachusetts, area. Recall that cysts, incubated at low and high temperatures, will excyst when exposed to an increase and decrease, respectively, in temperature. As the water warms during spring, overwintering benthic hypnozygotes excyst to seed the bloom. Field and laboratory studies suggest that the temperature increase is a primary trigger for excystment. The bloom

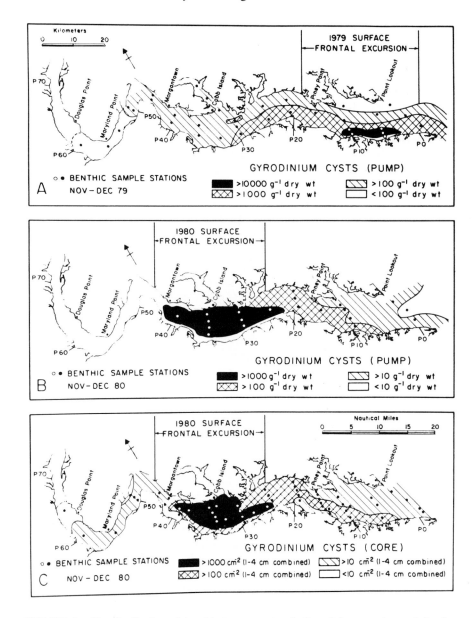

FIGURE 5.   The distribution of benthic hypnozygote of *Gyrodinium uncatenum* following blooms in 1979 and 1980. Highest concentrations of cysts reflect the position and movement of the front in both years. (From Tyler et al., *Mar. Ecol. Prog. Ser.* 7, 163, 1982. With permission.)

develops and continues through the summer. Population numbers remain high within the estuary due to limited tidal advection and limited numbers of cells in surface waters. In addition, the natural chelation of toxic metal ions by the high organic content of estuarine waters allows a higher viability of excysted cells than in coastal waters. Watras et al.[60] consider temperature and salinity the most important determinants of growth rate and peak abundance for vegetative *P. tamarensis* cells in estuaries. At some point during the bloom, despite apparently sufficient nutrients and optimum physical conditions, population numbers decline. Encystment, grazing, tidal advection, or toxicity may contribute to the decline. Watras et al.[60] found cell size decreased with rapid division at higher temperatures and speculated grazing may increase in importance in bloom decline when temperature exceeds 13°C.

Cysts from the spring-summer bloom accumulate in the sediments during the bloom. Cysts formed in the early part of the bloom may excyst to seed a short lived fall bloom when temperatures decrease. Cysts formed later in the bloom cannot break dormancy and overwinter until the spring water temperature increase triggers excystment.

In offshore waters, the picture is less complete. Deeper waters are less amenable to data collection than shallow estuaries of limited size. In the Gulf of Maine, the combination of extensive offshore cyst accumulations and upwelling and vertical mixing provide a mechanism for bloom initiation.[56,59] During upwelling periods cysts are carried in subsurface waters to the surface and are dispersed along the coast by horizontal flow. Excystment occurs at some time in surface waters. Blooms are supported and maintained when conditions of nutrients, temperature, salinity and other factors are suitable. Hypnozygotes form and drop out of the plankton and offshore currents return these cysts to seed beds.

Prior to 1972, blooms of *Protogonyaulax tamarensis* and associated PSP shellfish bed closures were common in the Bay of Fundy and along the eastern coast of Maine but records of *P. tamarensis* were rare and PSP unrecorded in the western Gulf of Maine and areas south. In fall, 1972, a massive bloom affected Maine, New Hampshire, and Massachusetts, and the shellfish beds were closed. The north-easterly winds preceding Hurricane Carrie have been credited with pushing surface waters containing *P. tamarensis* from the eastern Gulf of Maine southward along the coast.[56] Shellfish were contaminated as far south as Cape Cod during this outbreak. Cysts from this bloom seeded estuaries and offshore areas from Cape Cod north. In the estuaries around Cape Cod where tidal flushing is minimal, *P. tamarensis* populations have established themselves and blooms occur when conditions are suitable.[57,58] In other areas, such as Cape Ann, estuarine sampling recorded relatively low numbers of cysts, yet blooms still occur. These blooms are probably seeded by upwelled cysts from offshore seed beds.[56,57,61] Anderson et al.[57] also recorded viable cysts in estuarine sediments as far south as Connecticut and Long Island, New York, and advised monitoring of these areas for possible future PSP problems.

Whether the establishment of *P. tamarensis* populations south of Maine truly represents a southward extension of its range is unknown. Regardless, the success of the dispersal and survival of *P. tamarensis* populations south of its pre-1972 range is directly attributable to its benthic hypnozygote. The 1972 bloom introduced motile cells and the resulting cysts into unaffected areas. In favorable areas, these cysts established persistent populations.[57] In estuaries, the cysts seed overlying waters; in offshore areas cysts are upwelled and can provide inocula to distant bloom areas and may provide colonizers for new areas.

Interrelationships between biological and hydrographical factors can be seen in the three dinoflagellate species discussed above. Dinoflagellates do not exist in a static state in a static environment. They respond behaviorally and physiologically to changes in their environment. Responses have been optimized through selection for population survival and dispersal/accumulation. Responses to large scale changes in water mass characteristics (tidal exchange, frontal convergences, upwelling, vertical mixing) are receiving increased attention. Holligan[62] noted that blooms of *Gyrodinium aureolum* in tidal front regions off the British Isles appeared to be determined by physical and chemical factors in the seasonal thermocline that affected growth. Tyler and Seliger[31] described the physiological characteristics of *Prorocentrum minimum* that allowed subsurface transport from high salinity waters into low salinity waters and allowed bloom development. Dugdale[63] discussed the possible relationships between upwelling, nutrient utilization and interspecific competition in blooms of *Gonyaulax polyedra* off

Baja California. Steidinger and Haddad[15] discussed a mechanism of bloom formation in *Ptychodiscus brevis* including the coupling of oceanic frontal intrusion onto a shallow shelf and the sexual cycle.

The concentration of bloom dinoflagellates by hydrographic processes can be associated with sexual reproduction. Sexual reproduction may be triggered or enhanced by the physiochemical characteristics of these hydrographic processes or by the "conditioning" effect of large numbers of dinoflagellates (ectocrines, vitamins, etc.) in a discrete water mass or frontal system. Sexual reproduction may occur during non-bloom periods throughout a species' geographic range. However, keying sexual reproduction to bloom conditions increases gametic variability, the frequency of gametic fusion and genetic mixing. Whether nonbloom species respond to hydrographic processes and aggregate on a smaller scale for sexual reproduction is unknown. It is difficult to assess the frequency of sexual reproduction in natural populations of non-bloom species with present sampling techniques. It also is difficult to assess the impact of sexual reproduction on the genetic diversity and population dynamics of bloom and nonbloom species. The selection pressure towards either genetic diversity or fixation may vary with species. The lack of or low incidence of sexual reproduction observed in marine species may not be entirely due to limited sampling or inadequate culture techniques. Sexual reproduction rates may be low in some species in order to fix favorable genomes or to avoid loss of genetic material as in the case of benthic hypnozygotes in oceanic species. Dinoflagellates are a diverse and geologically old group. It is not unreasonable to assume a variety of genetic strategies are present in response to diverse habitats and modes of existence.

ACKNOWLEDGMENTS

I would like to thank Dr. Karen A. Steidinger and David Crewz for invaluable suggestions and criticisms during the many discussion sessions throughout the writing of this chapter. I also would like to thank Douglas Walker for his photography and dark-room work on the figures.

REFERENCES

1. Walker, L. M., Evidence for a sexual cycle in the Florida red tide dinoflagellate, *Ptychodiscus brevis* (= *Gymnodinum breve*). *Trans. Am. Microsc. Soc.,* 101, 287, 1982.
2. Loeblich, L. A. and Loeblich, A. R., III, The organism causing New England Red Tides: *Gonyaulax excavata,* in *Proceedings of the First International Conference on Toxic Dinoflagellate Blooms,* LoCicero, V. R., Ed., Mass. Sci. Tech. Found., Wakefield, MA, 1975, 207.
3. Steidinger, K. A., A re-evaluation of toxic dinoflagellate biology and ecology, in *Progress in Phycological Research,* Vol. II, Round, F. and Chapman, D., Eds., Elsevier North Holland, New York, 1983, 147.
4. Steidinger, K. A. and Williams, J., Dinoflagellates. Memoirs of the Hourglass Cruises, *Fla. Dep. Nat. Resour. Mar. Res. Lab.,* 2, 1970.
5. Elbrachter, M. and Drebes, G., Life cycles, phylogeny and taxonomy of *Dissodinium* and *Pyrocystis* (Dinophyta), *Helgol. Wiss. Meeresunters.,* 31, 347, 1978.
6. Wall, D. and Evitt, W. R., A comparison of the modern genus *Ceratium* Schrank, 1793, with certain Cretaceous marine dinoflagellates, *Micropaleontology,* 21, 14, 1976.
7. Kofoid, C. A. and Swezy, O., Free-living unarmored Dinoflagellata, *Mem. Univ. Calif.,* 5, 1, 1921.
8. Taylor, F. J. R., Dinoflagellates from the International Indian Ocean Expedition, *Biblio. Bot.,* 132, 1, 1976.
9. McLaughlin, J. J. A. and Zahl, P. A., Endozoic algae, in *Symbiosis,* Henry, S. M., Ed., Academic Press, New York, 1966, 257.

10. Pfiester, L. A. and Lynch, R. A., Amoeboid stages and sexual reproduction of *Cystodinium bataviense* and its similarity to *Dinococcus* (Dinophyceae), *Phycologia,* 19, 178, 1980.
11. Omori, M. and Hamner, W. M., Patchy distribution of zooplankton: Behavior, population assessment and sampling problems, *Mar. Biol.,* 72, 193, 1982.
12. Zingmark, R. G., Sexual reproduction in the dinoflagellate *Noctiluca miliaris* Suriray, *J. Phycol.,* 6, 122, 1970.
13. Anderson, D. M. and Wall, D., Potential importance of benthic cysts of *Gonyaulax tamarensis* and *G. excavata* in initiating toxic dinoflagellate blooms, *J. Phycol.,* 14, 244, 1978.
14. Anderson, D. M., Effects of temperature conditioning on development and germination of *Gonyaulax tamarensis* (Dinophyceae) hypnozygotes, *J. Phycol.,* 16, 166, 1980.
15. Steidinger, K. A. and Haddad, K., Biologic and hydrographic aspects of red tides, *BioScience,* 31, 814, 1982.
16. Tyler, M. A., Coats, D. W., and Anderson, D. M., Encystment in a dynamic environment: Deposition of dinoflagellate cysts by a frontal convergence, *Mar. Ecol. Prog. Ser.,* 7, 163, 1982.
17. Buchanan, R. J., Studies at Oyster Bay in Jamaica, West Indies. IV. Observations on the morphology and asexual cycle of *Peridinium bahamense* Plate, *J. Phycol.,* 4, 272, 1968.
18. Braarud, T., Observations on *Peridinium trochoideum* (Stein) Lemm. in culture, *Nytt Mag. Bot.,* 6, 39, 1957.
19. Pfiester, L. A., Sexual reproduction of *Peridinium cinctum f. ovoplanum* (Dinophyceae), *J. Phycol.,* 11, 259, 1975.
20. Pfiester, L. A., Sexual reproduction of *Peridinium willei* (Dinophyceae), *J. Phycol.,* 12, 234, 1976.
21. Pfiester, L. A., Sexual reproduction of *Peridinium gatunense* (Dinophyceae), *J. Phycol.,* 13, 92, 1977.
22. Silva, E. S., Some observations on marine dinoflagellates cultures. II. *Glenodinium foliaceum* Stein and *Gonyaulax diacantha* (Meun.) Schiller, *Bot. Mar.,* 3, 75, 1962.
23. Swift, E. and Wall, D., Asexual reproduction through a thecate stage in *Pyrocystis acuta* Kofoid, 1907 (Dinophyceae), *Phycologia,* 11, 57, 1972.
24. Weiler, C. S. and Eppley, R. W., Temporal patterns of division in the dinoflagellate genus *Ceratium* and its application to the determination of growth rate, *J. Exp. Mar. Biol. Ecol.,* 39, 1, 1979.
25. Coats, D. W. and Heinbokel, J. F., A study of reproduction and other life cycle phenomena in planktonic protists using an acridine orange fluorescence technique, *Mar. Biol.,* 67, 71, 1982.
26. Sweeney, B. M. and Hastings, J. W., Rhythmic cell division in populations of *Gonyaulax polyedra,* *J. Protozool.,* 5, 217, 1958.
27. Hastings, J. W. and Sweeney, B. M., Phased cell division in the marine dinoflagellates, in *Synchrony in cell division and growth,* Zeuthen, E., Ed., John Wiley and Sons, New York, 1964, 307.
28. Weiler, C. S., Population structure and in situ division rates of *Ceratium* in oligotrophic waters of the North Pacific central gyre, *Limnol. Oceanogr.,* 25, 610, 1980.
29. Swift, E., Stuart, M., and Meunier, V., The in situ growth rates of some deep-living oceanic dinoflagellates: *Pyrocystis fusiformis* and *Pyrocystis noctiluca, Limnol. Oceanogr.,* 21, 418, 1976.
30. Seliger, H. H., Carpenter, J. H., Loftus, M., and McElroy, W. D., Mechanisms for the accumulation of high concentrations of dinoflagellates in a bioluminescent bay, *Limnol. Oceanogr.,* 15, 234, 1970.
31. Tyler, M. A. and Seliger, H. H., Selection for a red tide organism: Physiological responses to the physical environment, *Limnol. Oceanogr.,* 26, 310, 1981.
32. Anderson, D. M., Chisholm, S. W., and Watras, C. J., The importance of life cycle events in the population dynamics of *Gonyaulax tamarensis, Mar. Biol.,* in press.
33. Schmitter, R. E., Temporary cysts of *Gonyaulax excavata*: Effects of temperature and light, in *Toxic Dinoflagellate Blooms,* Taylor, D. L. and Seliger, H. H., Eds., Elsevier North Holland, New York, 1979, 123.
34. Chapman, D. V., Livingstone, D., and Dodge, J. D., An electron microscope study of the encystment and early development of the dinoflagellate *Ceratium hirundinella* (Dinophyceae), *Br. Phycol. J.,* 16, 183, 1981.
35. Chapman, D. V., Dodge, J. D., and Heaney, S. I., Cyst formation in the freshwater dinoflagellate *Ceratium hirundinella* (Dinophyceae), *J. Phycol.,* 18, 121, 1982.
36. Cao Vien, De. M., Sur l'existence de phénomènes sexuels chez un Péridinien libre, l'*Amphidinium Carteri, C. R. Acad. Sci. Ser. D.,* 264, 1006, 1967.
37. Von Stosch, H. A., La signification cytologique de la 'cyclose nucléaire' dans le cycle de vie des Dinoflagelles, *Soc. Bot. Fr., Memoires,* 201, 1972.
38. Beam, C. A. and Himes, M., Evidence for sexual fusion and recombination in the dinoflagellate *Crypthecodinium (Gyrodinium) cohnii, Nature,* 250, 435, 1974.
39. Taylor, F. J. R., A description of the benthic dinoflagellate associated with maitotoxin and ciguatoxin, including observations on Hawaiian material, in *Toxic Dinoflagellate Blooms,* Taylor, D. L. and Seliger, H. H., Eds., Elsevier North Holland, New York, 1979, 71.

40. Diwald, K., Die ungeschlechtliche und geschlechtliche Fortpflanzung von *Glenodinium lubiniensiforme* spec. nov., *Flora,* Jena, 32, 174, 1937.

41. Walker, L. M. and Steidinger, K. A., Sexual reproduction in the toxic dinoflagellate *Gonyaulax monilata, J. Phycol.,* 15, 312, 1979.

42. Dale, B., Cysts of the toxic red-tide dinoflagellate *Gonyaulax excavata* (Braarud) Balech from Osolofjorden, Norway, *Sarsia,* 63, 29, 1977.

43. Dale, B., Yentch, C. M., and Hurst, J., Toxicity in resting cysts of the red tide dinoflagellate *Gonyaulax excavata* from deeper water coastal sediments, *Science,* 201, 1223, 1978.

44. Turpin, D. H., Dobell, P. E. R., and Taylor, F. J. R., Sexuality and cyst formation in Pacific strains of the toxic dinoflagellate *Gonyaulax tamarensis, J. Phycol.,* 14, 235, 1978.

45. Spero, H. J. and Morée, M. D., Phagotrophic feeding and its importance to the life cycle of the holozoic dinoflagellate, *Gymnodinium fungiforme, J. Phycol.,* 17, 43, 1981.

46. Von Stosch, H. A., Observations on vegetative reproduction and sexual life cycles of two freshwater dinoflagellates, *Gymnodinium pseudopalustre* Schiller and *Woloszynskia apiculata* sp. nov., *Br. Phycol. J.,* 8, 105, 1973.

47. Pfiester, L. A. and Skvarla, J. J., Heterothallism and thecal development in the sexual life history of *Peridinium volzii* (Dinophyceae), *Phycologia,* 18, 13, 1979.

48. Yoshimatsu, S., Sexual reproduction of *Protogonyaulax catenella* in culture. I. Heterothallism., *Bull. Plankton Soc. Jpn.,* 28, 131, 1981.

49. Taylor, D. L., The cellular interactions of algae-invertebrate symbiosis, in *Advances in Marine Biology,* Vol. 11, Russell, F. S. and Yonge, M., Eds., Academic Press, New York, 1973, 1.

50. Bernstein, H., Recombinational repair may be an important function of sexual reproduction, *BioScience,* 33, 326, 1983.

51. Dale, B., Dinoflagellate resting cysts: "benthic plankton", in *Survival Strategies in the Algae,* Fryxell, G. A., Ed., Cambridge University Press, 1983, 69.

52. Trainor, F. R., Is a reduced level of nitrogen essential for *Chlamydomonas eugametos* mating in nature?, *Phycologia,* 14, 167, 1975.

53. Cain, J. R., Inhibition of zygote germination in *Chlamydomonas moewusii* (Chlorophyceae, Volvocales) by nitrogen deficiency and sodium citrate, *Phycologia,* 19, 184, 1980.

54. Von Stosch, H. A., Zum Problem der sexuellen fortpflanzung in der Peridineengattung *Ceratium, Helgol. Wiss. Meeresunters.,* 10, 140, 164.

55. Prakash, A., Growth and toxicity of a marine dinoflagellate, *Gonyaulax tamarensis, J. Fish. Res. Board Can.,* 24, 1589, 1967.

56. Hartwell, A. D., Hydrographic factors affecting the distribution and movement of toxic dinoflagellates in the western Gulf of Maine, in *Proceedings of the First International Conference on Toxic Dinoflagellate Blooms,* LoCicero, V. R., Ed., Mass. Sci. Tech. Found., Wakefield, MA, 1975, 47.

57. Anderson, D. M., Kulis, D. M., Orphanos, J. A., and Ceurvels, A. R., Distribution of the toxic dinoflagellate *Gonyaulax tamarensis* in the southern New England region, *Estuarine Coastal Shelf Sci.,* 14, 447, 1982.

58. Anderson, D. M. and Morel, F. M. M., The seeding of two red tide blooms by the germination of benthic *Gonyaulax tamarensis* hypnocysts, *Estuarine Coastal Mar. Sci.,* 8, 279, 1979.

59. Lewis, C. M., Yentsch, C. M., and Dale, B., Distribution of *Gonyaulax excavata* resting cysts in sediments of Gulf of Maine, in *Toxic Dinoflagellate Blooms,* Taylor, D. L. and Seliger, H. H., Eds., Elsevier North Holland, New York, 1979, 235.

60. Watras, C. J., Chisholm, S. W., and Anderson, D. M., Regulation of growth in an estuarine clone of *Gonyaulax tamarensis* Lebour: Salinity-dependent temperature responses, *J. Exp. Mar. Biol. Ecol.,* 62, 25, 1982.

61. Mulligan, H. F., Oceanographic factors associated with New England red tide blooms, in *Proceedings of the First International Conference of Toxic Dinoflagellate Blooms,* LoCicero, V. R., Ed., Mass. Sci. Technol. Found., Wakefield, MA, 1975, 23.

62. Holligan, P. M., Dinoflagellate blooms associated with tidal fronts around the British Isles, in *Toxic Dinoflagellate Blooms,* Taylor, D. L. and Seliger, H. H., Eds., Elsevier North Holland, New York, 1979, 249.

63. Dugdale, R. C., Primary nutrients and red tides in upwelling regions, in *Toxic Dinoflagellate Blooms,* Taylor, D. L. and Seliger, H. H., Eds., Elsevier North Holland, New York, 1979, 257.

Chapter 3

CELLULAR SPECIALIZATION AND REPRODUCTION IN PLANKTONIC FORAMINIFERA AND RADIOLARIA

O. Roger Anderson

TABLE OF CONTENTS

I. INTRODUCTION

Planktonic foraminifera and radiolaria are holoplanktonic protozoa widely distributed in the open ocean. They possess a fine network of peripheral rhizopodia, or also in some radiolaria a corona of radially-arranged, ray-like axopodia and attached vacuolated cytoplasm, whereby they capture prey and perhaps enhance their buoyancy (Figures 1, 2). Much of our basic knowledge about the gross morphology and fundamental physiology of planktonic foraminifera[1,2] and radiolaria[3-6] can be traced to studies in the late nineteenth and early twentieth centuries. Recent investigations employing electron microscopy and modern physiological research methods have contributed finer details of their cytology,[7-16] reproduction, symbiotic relationships, nutrition, feeding behavior, and other basic physiological processes.[17-35] Consequently, we are beginning to better understand the role of planktonic foraminifera and radiolaria in marine plankton communities. An analysis of natural prey snared in the rhizopodia of freshly collected specimens from the natural environment and experimental organisms cultured in the laboratory, while offered algal and animal prey, indicates that a wide variety of phytoplankton and microzooplankton are consumed by planktonic foraminifera and radiolaria.[26-29] In some species, the rhizopodia are connected with a frothy envelope of bubble-like alveoli surrounding the central cell mass thus enhancing buoyancy as occurs in the planktonic foraminifer *Hastigerina pelagica*, Figure 1, and in radiolaria (e.g., *Thalassicolla* sp., a solitary form, Figure 2; and *Sphaerozoum punctatum*, Müller, Figure 4, a colonial species).

The comparative morphology of some living specimens of planktonic foraminifera and radiolaria is presented in Figures 1 to 4. Planktonic foraminifera occur as nonspinose and spinose species. The latter group, as shown in Figures 1 and 3, bear long spines attached to their shells that support the rhizopodia, whereas the nonspinose forms are surrounded by a self-supported web of rhizopodia. Radiolaria are classified in three major groups: (1) Spumellaria with nearly spherical cell bodies surrounded by a perforated wall and sometimes enclosed by a complex lattice shell composed of silica (Figure 2 and inset, Figure 8); (2) Nassellaria with an ovoid to spheroidial cell body containing pores only at one end of the cell body (Figure 7), and sometimes surrounded by a delicate helmet-shaped skeleton; and (3) Phaeodaria (not illustrated) with an assembly of three complex pore fields on the surface of the cell body, a distinctive dense mass of pigmented debris in the rhizopodia, and a variety of shell morphologies varying from cup-shaped to complex geodesic lattices. All skeletons in radiolaria are composed of amorphous silica. Much of the research reported here concerns the Spumellaria which have been studied in the laboratory much more extensively than Nassellaria or Phaeodaria. Recent publications on the fine structure and biology of planktonic foraminifera,[30-32] and radiolaria[33-37] provide general background information; therefore, only relevant concepts of cellular specialization are presented here as a context for a more detailed discussion of reproductive processes and life history.

II. PATTERNS OF CELLULAR SPECIALIZATION

Among the marine Sarcodina, planktonic foraminifera and radiolaria exhibit microscopic evidence of advanced cellular specialization in comparison to more primitive forms such as the amoebae. Electron microscopic evidence of increasing cellular specialization within the three groups is presented through comparative electron micrographs of a marine amoeba, *Hartmannella* sp. (Figure 5),[38] a planktonic foraminifer, *Orbulina universa* (Figure 6), and radiolaria (nassellarian species, Figure 7, and a spumellarian, *Spongodrymus* sp., Figure 8).

Amoebae possess a fluid and highly mobile cytoplasmic organization as revealed by

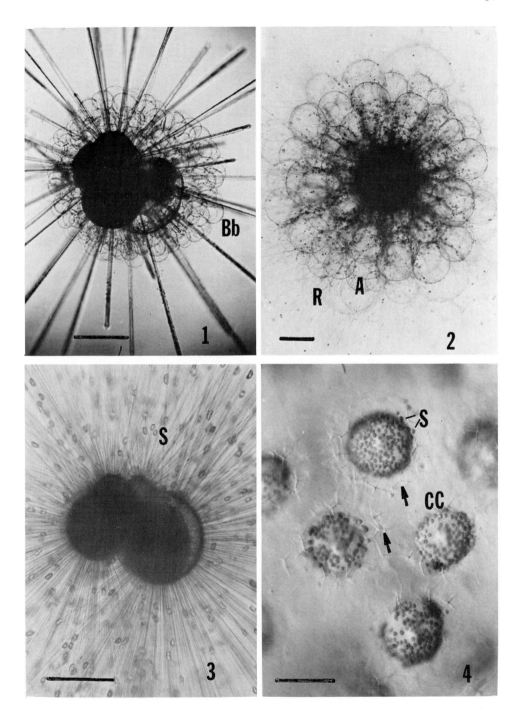

FIGURES 1 to 4.   Comparative views of living planktonic foraminifera and radiolaria. Figure 1, *Hastig-erina pelagica*, a spinose planktonic foraminifera possessing a distinctive bubble-like mass (Bb) of alveolate cytoplasm surrounding the spiral calcite shell; marker = 500 μm. Figure 2,[26] a skeletonless radiolarian *Thal-assicolla nucleata* with a dense spherical central capsule surrounded by alveoli penetrated by rhizopodia (R) bearing symbionts; marker = 500 μm. Figure 3, *Orbulina universa* (spiral stage) a spinose planktonic fora-minifera bearing numerous symbionts (S) on its spines; marker = 250 μm. Figure 4,[45] a small portion of a colonial radiolarian *Sphaerozoum punctatum* containing central capsules (CC) interconnected by rhizopo-dia, enclosing dinoflagellate symbionts (S) and symmetrical, triradiate siliceous spicules (arrows); marker = 250 μm.

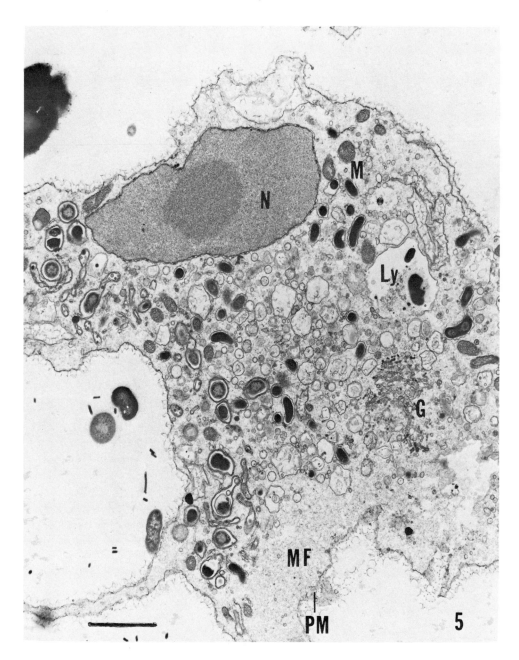

FIGURE 5.   Fine structure of a marine amoeba *Hartmannella* sp. exhibits the significant features of diffuse cellular specialization, including the dispersed distribution of subcellular organelles: nucleus (N), Golgi (G), lysosomes (Ly), Mitochondria (M), microfilaments (MF) near the plasma membrane (PM), and vacuoles containing densely-stained, possibly symbiotic bacteria distributed throughout the cytoplasm. Although some of the organelles may be occasionally assembled in groups, there are no large specialized zones segregated by membranous or non-living boundaries. Marker = 2 μm. (From Anderson, O. R., *J. Protozool.*, 24, 370, 1977. With permission.)

the massive protoplasmic flow accompanying locomotion when viewed with the light microscope. Their fine structure also exhibits a dispersed organization of subcellular organelles without substantial membranous or non-living boundaries to compartmentalize regions of the cell. A fine layer of microfilaments (MF, Figure 5) occurs at the

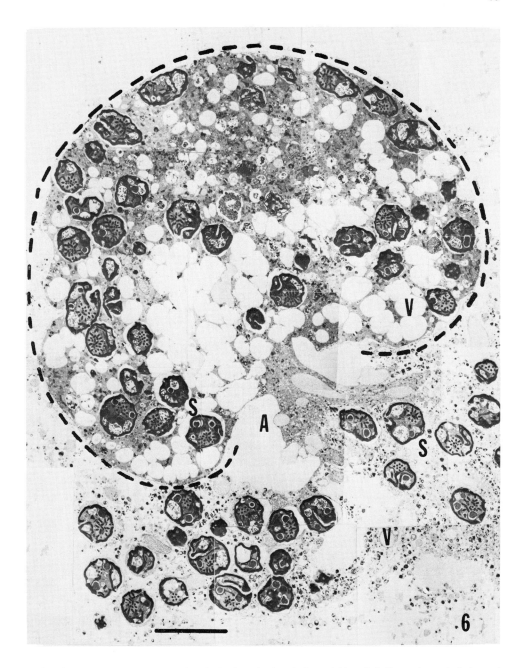

FIGURE 6. A montage of a section through a planktonic foraminifer, *Orbulina universa* (spiral stage), displays a transitional pattern of specialization including the final chamber surrounded by a calcite wall (represented by dashed lines as the wall was decalcified before sectioning), a graded continuity between intrashell and extrashell cytoplasm connecting through the aperature (A), and the distribution of symbionts (S), and many vacuoles (V) occurring in the intrashell and to a lesser extent in the extrashell cytoplasm. Marker = 30 μm.

periphery of the cell beneath the plasma membrane and extends into the pseudopodia. However, microfilaments are not segregated by a membranous boundary, but gradually merge with the granular cytoplasm containing the membranous organelles in the cortical regions of the cell. Much of the cytoplasm exhibits a diffuse pattern of orga-

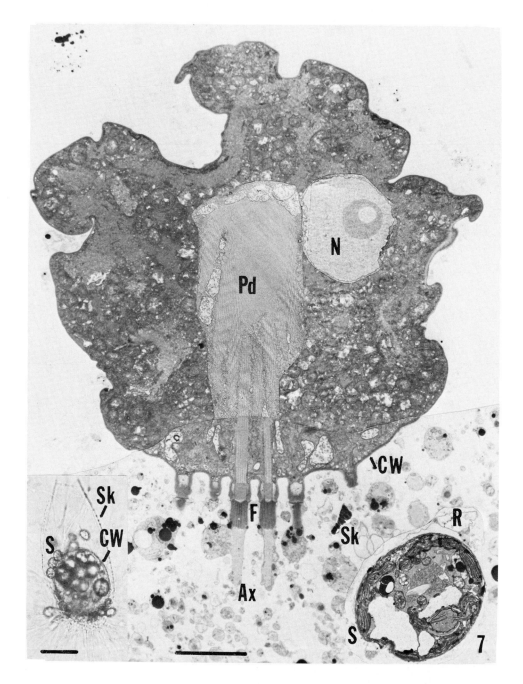

FIGURE 7.   A composite electronmicrograph of a nassellarian radiolarian displays the nucleus (N), intracapsular microtubules arranged as a cone called a podoconus (Pd) surrounded by cytoplasm enclosed within the capsular wall (CW) penetrated by fusules (F). Fusules are cytoplasmic strands with internal shafts of microtubules, connecting intracapsular cytoplasm with extracapsular cytoplasm containing the axopodia (Ax), that have been contracted during fixation, vacuolated digestive matrix and symbionts (S) enclosed within rhizopodia. The clearly demarked zones of specialized function are characteristic of zonal specialized organisms; marker = 5 μm. Inset, light micrograph of a living specimen surrounded by a characteristic siliceous shell called a cephalis; marker = 45 μm. Only fragments of the siliceous cephalis (Sk) are present in the electron micrograph. The remainder of the larger globose skeleton surrounding the central capsule is not included within this view.

41

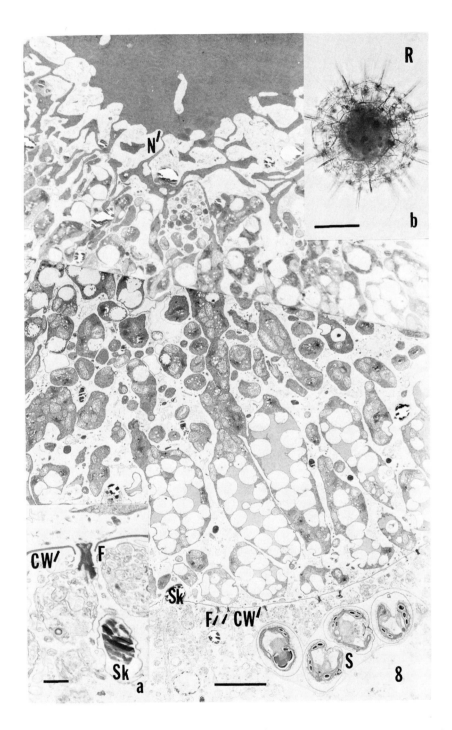

FIGURE 8.   A composite fine structure view of a spumellarian radiolarian, *Spongodrymus* sp., further illustrates the major compartments of a zonal specialized organism including the spherical capsular wall (CW) surrounding the lobate intracapsulum, and penetrated by strands of cytoplasm forming the numerous fusules (F) scattered over the entire surface of the capsular wall, the extracapsulum containing vacuoles, the cytokalymma, a cytoplasmic envelope surrounding a portion of the siliceous skeleton (Sk), and rhizopodia with attached symbionts (S); marker = 10 μm. Inset (a), lower left, is an enlarged view of the capsular wall exhibiting the fusule (F) and nearby segment of skeleton (Sk) enclosed by the cytokalymma, marker = 1 μm. Inset (b), is a light micrograph of a skeleton-bearing radiolarian possessing an elaborate lattice shell penetrated by numerous rhizopodia (R) emerging from a dense central capsule; marker = 100 μm.

nelle distribution with only limited localized differentiation. Protozoa with this pattern of organization, lacking distinct major compartmentalized regions of specialization, will be classified as "diffuse specialized."

In contrast to the amoebae, with only limited protective covering largely on the cell surface, planktonic foraminifera possess a multi-chambered calcite shell forming a non-living barrier that partially separates the dense intrashell cytoplasm from the more frothy or web-like extrashell cytoplasm containing the rhizopodia and a vacuolated digestive matrix. Each chamber is connected to its neighbor by a small aperture that permits continuity of the cytoplasm within the shell. An external aperture in the final, largest chamber (A, Figure 6) permits continuity between intrashell and extrashell cytoplasm. The cytoplasmic transition, however, is not abrupt; the intrashell cytoplasm gradually merges into the extrashell cytoplasm. Moreover, the position of the transition region near the aperture may vary among species or according to the physiological state of the foraminifer. When the foraminifer is well nourished and the intrashell cytoplasm is abundant, the dense cytoplasm may extend out of the aperture and forms a thin envelope surrounding the shell. The rhizopodia and external vacuolate matrix emerge from this thin cytoplasmic envelope. Algal symbionts, when present, are enclosed in perialgal vacuoles distributed throughout the intrashell and extrashell cytoplasm. However, there is clear evidence of specialized activity in regulating the position of the symbionts. Many are distributed in the peripheral rhizopodial network outside of the shell during the day and gathered into the central cytoplasm at night. Digestive vacuoles are observed in both the intrashell and extrashell cytoplasm.[24,28] Food vacuoles may also be observed within the shell when the foraminifer has consumed large prey and massive amounts of food particles are engulfed and carried into the shell for subsequent digestion. The perforated calcite shell wall is an extracellular product secreted by a living membranous template assembled shortly before the wall is deposited.[39] The wall apparently is not intimately associated with the living substance of the cell after it has been deposited. Other than the complex calcite shell wall, there are few permanent structures that partition large regions of the cytoplasm for specialized functions. The frothy bubble-like capsule in *Hastigerina pelagica* is an interesting exception. It is a relatively stable specialized cytoplasmic region providing enhanced buoyancy. Other planktonic foraminifera, possess large intracellular vacuoles, called fibrillar bodies,[24] filled with a puffy organic matter that may also aid buoyancy. In overview, the foraminifera exhibit structural adaptations to enhance cellular compartmentalization and specialization that are more stable and complex than observed in amoebae, particularly with reference to the foraminiferal shell wall. The distribution of cytoplasmic organelles in planktonic foraminifera as in amoebae is somewhat diffuse. The foraminifera, however, exhibit evidence of increasing regional specialization as between intrashell and extra-shell cytoplasm. Thus, planktonic foraminifera are classified here as "transitional specialized protozoa". Their degree of cytoplasmic specialization is at an intermediate level compared to lower forms such as amoebae and to more complex forms, as in radiolaria.

Radiolarian cytoplasm is clearly compartmentalized into two major regions: (1) a central capsule,[6] surrounded by a perforated capsular wall, containing the nucleus or nuclei in multinucleated species, membranous organelles, and food reserves, and (2) the extracapsulum[6] composed of the rhizopodia, frothy or alveolated cytoplasm containing digestive vacuoles, and algal symbionts, when present, enclosed within perialgal vacuoles or loosely adhering to the rhizopodial strands. The capsular wall (CW, Figures 7 and 8) is a permanent structure in all known radiolaria and consists of a living membranous boundary, sometimes containing a thickened non-living organic deposit that forms a distinct barrier (the wall proper), perforated with strands of cytoplasm connecting the intracapsulum to the extracapsulum. The strands of cytoplasm, called fusules[10] (F, Figures 7 and 8), are complicated structures varying in intricacy among species. Their fine structure and mode of connection with the capsular wall membrane varies among taxa. The Nassellaria for example possess cylindrical fusules localized at one pole of the prolate central capsule (Figure 7). The fusules contain shafts of microtubules extending from a large conical microtubule-organizing-center within the central capsule. The Spumellaria have a nearly spherical central capsule perforated on the surface with numerous fusules. In some cases the fusule strand passes through a collar-like opening[23] (inset a, Figure 8) which may permit continuity between the seawater surrounding the central capsule and the free space surrounding intracapsular lobes. This lobate arrangement is common in the large spumellarian radiolaria possessing a spongiose skeleton. Each lobe in the intracapsulum (Figure 8) is further differentiated in some species into a distal segment (containing vacuoles and reserve substances) and a proximal segment rich in mitochondria, Golgi apparatus, vesicles, and endoplasmic reticulum. Symbionts have not been reported within the central capsule. This is not surprising as the symbiont is too large to pass through the fusule. The skeleton when present is composed of amorphous silica secreted within an elaborate cytoplasmic structure called a cytokalymma[23,36,40] which persists after skeletal production as a thin cytoplasmic sheath surrounding the skeletal elements. The cytokalymma[23,35] is a living template that precedes silica deposition and establishes the architecture of the skeleton, an elaborate and sometimes ornate structure, surrounding or also included within the central capsulum. Recent biochemical evidence has been obtained in our laboratory to further substantiate the functional specialization of the intracapsulum compared to the extracapsulum in two species of a large solitary radiolarian, *Thalassicolla*. The activity of acid aryl phosphatase, a marker enzyme for digestive activity, was significantly greater in the extracapsulum as compared to the intracapsulum, whereas the activity of cytochrome oxidase, a marker enzyme for aerobic activity, was significantly higher in the intracapsulum compared to the extracapsulum. The markedly increased activity of acid aryl phosphatase in the extracapsulum compared to the intracapsulum is of interest since the Golgi apparatus, the site of lysosome production and presumably therefore the major source of acid aryl phosphatase, is in the intracapsulum. Hence the lysosomes containing acid aryl phosphatase must be transported through the fusule into the digestive zone of the extracapsulum. The digestive zone, moreover, in most radiolaria examined with the electron microscope is located proximal to the capsular wall. Given the rather stable specialized zones in radiolaria, they are classified as *"zonal specialized"* organisms.

The marked zonal specialization of radiolaria compared to planktonic foraminifera is partially due to: (1) the distinctive elaborate capsular wall that effectively segregates intracapsulum from extracapsulum, (2) the intricate fusules connecting the intracapsulum with the extracapsulum, (3) the localization of the digestive activity in a definite zone near the capsular wall rather than being dispersed within the cytoplasm inside the central cell body and outside of it as occurs in planktonic foraminifera, and (4) the persistent cytokalymma in radiolaria that secretes the elaborate siliceous skeleton. The symbionts in radiolaria, moreover, are localized in the peripheral cytoplasm or gathered in close to the capsular wall but are not taken into the central capsule. The symbionts in planktonic foraminifera and radiolaria are moved by cytoplasmic streaming along the host rhizopodial network, and therefore, their position is determined by the host. A daily rhythm of symbiont dispersal into the peripheral rhizopodia during daylight and their withdrawal near the central cell body at night occurs in radiolaria as in planktonic foraminifera.[24] There appears, however, to be a much less intimate structural association between radiolaria and their symbionts as compared to the symbiont-bearing planktonic foraminifera (e.g., *Globigerinoides sacculifer*), where the algae are gathered into the smallest, earliest-formed chambers of the foraminifer.[24] It is not

Table 1

CATEGORIES OF CELLULAR SPECIALIZATION IN PROTOZOA

Diffuse	Transitional	Zonal
The cytoplasmic organelles are dispersed widely throughout the cell. There are no specialized large regions enclosed by stable compartmentalizing barriers such as living membranes or nonliving boundaries secreted by the cell. Regions of differentiation, when they occur, are often transitory, depending on the physiological state of the organism. They merge with the surrounding cytoplasmic regions rather than being distinctly separate (Figure 5).	Specialized regions of cytoplasm occur to varying degrees with stable, clearly identifiable organellar structures, but these regions merge with one another without permanent compartmentalizing barriers between the zones of specialization. Cellular secretory products may form perforated walls or boundaries partially segregating the zones from one another, but specialized connecting structures across the boundaries are rudimentary or lacking (Figure 6).	There are clearly identifiable, major differentiated zones of cytoplasm, often separated by stable boundaries such as compartmentalizing living membranes containing specialized structures mediating connections between the zones (Figures 7,8). Evidence for structural differentiation is further substantiated by physiological specialization within the zones.

possible to conclude at present whether the greater degree of zonal specialization in radiolaria indicates a more advanced stage of biological evolution, but it is an interesting hypothesis to be considered for further evaluation.

Table 1 presents the major characteristics of diffuse, transitional, and zonal specialized protozoa and is a summary of the main features to be used in subsequent sections of this chapter where reproductive processes are related to the degree of specialization exhibited by planktonic foraminifera and radiolaria. Although this classificatory scheme can be applied to a broad range of protozoa, it will be limited here to pertinent members of the Sarcodina.

III. MATURATION AND REPRODUCTION

Our knowledge of the life cycles for planktonic foraminifera and radiolaria is incomplete and restricted largely to an understanding of events during maturation, cellular differentiation prior to reproduction, and release of reproductive swarmers. The term reproductive swarmer will be used as a general category for flagellated cells released at reproductive maturity and will include asexual cells (zoospores) and sexual reproductive cells (gametes). Where evidence indicates that the swarmers are gametes, as in planktonic foraminifera, the basis for the conclusion that they are gametes will be given. Little is known about the events following release of reproductive swarmers or the very earliest stages of the new generation. This is attributed in part to limitations in laboratory culture that apparently do not offer a suitable environment for prolonged survival or continued development of the reproductive cells after they are released. Moreover, the small size of the very early stages of development (often <30 μm) makes it difficult to observe them in the open ocean or to capture them without trauma. Nonetheless, recent laboratory-based studies on specimens maintained in short-term cultures and observational data on specimens obtained directly from the natural environment have elucidated some of the major events that occur during maturation and reproduction, and some of the ecological and physiological factors that may influence these events. A description of the major stages of maturation and cellular differentiation leading to swarmer release in planktonic foraminifera and radiolaria is presented as a context for discussion of environmental and physiological regulating factors presented in the next section.

A. Foraminifera

1. Maturation

The calcite shell of planktonic foraminifera is a multichambered structure that originates in the very earliest recognizable stages as a spherical or nearly spherical solitary shell (<30 μm) called the proloculus containing a small aperture permitting continuity between intrashell and extrashell cytoplasm. As the foraminifer matures, additional chambers of increasing size are secreted resulting in juvenile stages (approximately 30 to 100 μm in size) bearing the species-specific morphology of the mature multichambered shell. The shell grows by periodic addition of the chambers, and usually with concomitant thickening of the existing chamber walls. Hence the size of the shell is a good indicator of the progressive maturation of the specimen and its vitality. In addition to the increase in shell size and rate of development, the amount of extrashell cytoplasm and the streaming activity of the rhizopodia are also useful indicators of specimen vigor. When a newly formed chamber has been secreted, it often appears translucent and lacks substantial amounts of cytoplasm. However, with subsequent thickening of the calcite wall, the chamber becomes less translucent and if the specimen is well-nourished a dense deposit of cytoplasm fills the new chamber. In some cases, the intrashell cytoplasm emerges from the aperture as a vacuolated mass or as a layer reflexed over the outer surface of the shell. Variations in cytoplasmic abundance within the chambers and surrounding the aperture(s) occur among species of planktonic foraminifera. Species bearing spines, offer further evidence of vitality as unfavorable growth conditions frequently result in spine shortening or shedding, accompanied by loss of buoyancy. If the unfavorable conditions are reversible and the organism is returned to a supportive environment, the spines may be regenerated and floating resumed. Frequently, spinose species will not accept food when the spines are short or shed which suggests that the spines serve a significant enabling function in supporting rhizopodia that cover their surface and radiate out in the surrounding space thereby enhancing prey capture. Nonspinose species also capture prey in their fine rhizopodial network which is largely self-supported. The abundance and vigor of algal symbionts; when present, offer further evidence for vitality and maturity. Among the various categories of free-swimming and attached algal cells associated with some planktonic foraminifera,[24,30] the most useful in monitoring vigor and normal development of the host are the endosymbiotic, coccoid dinoflagellate stages, harbored in the rhizopodia and enclosed within the intrashell cytoplasm. The daily cycle of symbiont dispersal into the peripheral cytoplasm at daybreak and withdrawal into the central cytoplasm at night is a useful sign in monitoring vitality and assessing progress toward reproduction. As explained subsequently, some of the symbionts are shed in *G. sacculifer* during early stages of reproduction.

2. Reproduction

a. General Processes

An analysis of maturation and reproduction in three spinose planktonic foraminiferal species, *Globigerinella aequilateralis, Globigeriniodes sacculifer* and *Hastigerina pelagica* exhibits certain common features that can be generalized for the group in addition to specific variations to be presented hereafter.

One of the first signs of impending reproduction is prey rejection or lack of digestion of prey captured within the rhizopodia. An otherwise healthy robust specimen that has reached a mature size will usually accept crustacean prey (*Artemia* nauplii or copepods) each day. Rejection of prey is either a sign of reproduction or loss of vitality. If the general indicators of health (spine condition, cytoplasmic abundance, rhizopodial streaming) are positive, prey rejection is usually an indicator of early stages of reproduction. The next prominent morphological indication of reproduction is breaking and shedding of the spines leaving a characteristic scar in the shell where the spine was resorbed.[41,42] Simultaneously in *H. pelagica*, the cytoplasm becomes a bright scarlet or orange red, a culmination of a gradual intensification of red pigmentation commencing during early maturation of the organism. The bubble capsule (mass of alveolated cytoplasm surrounding the shell) is also resorbed in *H. pelagica*. The loss of spines and withdrawal of extrashell cytoplasm is accompanied by loss of buoyancy. The spine shedding and sinking process occur within a period of about twenty hours before the reproductive cells are released. After settling, the contracted cytoplasm surrounding the shell may become fuzzy in appearance (as in *G. sacculifer*) due to the many fine hair-like rhizopodia emanating from its surface. A bulge of cytoplasm, or sometimes two, appears at the aperture of the shell and expands outward eventually rupturing to release myriads of flagellated swarmers and/or producing long thin strands of beaded cytoplasm that yield additional swarmers. The free flagellated swarmers may hover near the shell in gyrating motions or swim off in many directions with an undulating, but rapid motion. In some cases masses of the developing swarmers may cling together, though released from the mother cytoplasm, until they become fully separated and swim away. Not all swarmers appear competent in laboratory cultured specimens and a goodly number may accumulate as a granular mass near the aperture of the shell. Based on light microscopic examination of thin sections from swarmer producing organisms, the number of swarmers was estimated to be approximately 3×10^5 per *G. sacculifer*.[41] Each swarmer is about 5 μm in diameter and possesses two flagella.

It is not possible to determine at present whether the swarmers are haploid or diploid; thus we cannot state conclusively that they are gametes. It is suspected, however, that they may be gametes, given their small size and the possibility of fusion.[41,43] The major events during swarmer production and release from *H. pelagica* are represented in Figures 9 to 14 and as a narrative outline in Table 2. Table 3 contains an outline of events leading to swarmer production in *G. sacculifer*.

In many respects, *H. pelagica* is unique compared to other planktonic foraminifera; its shell wall is thinner than most spinose planktonic species, achieving shell lengths of >750 μm along the longest axis and a total diameter of 2.0 μm including spines and bubble capsules (compared to more typical shell dimensions in other species of 400 to 700 μm) and a total diameter of 1 mm with spines. Moreover, the bubble capsule is a clearly differentiating structure not found in other foraminifera. It is not surprising, therefore, that fine details of swarmer production and release may differ among species of planktonic foraminifera with diverse morphologies and life histories. This becomes especially clear when ultra-thin sections of swarmer-producing cells are examined by electron microscopy.

b. Fine Structure

Comparative data for *Globigerinella aequilateralis*, *Globigerinoides sacculifer*, *G. ruber* and *Hastigerina pelagica* are presented. During the earliest stages of swarmer production, before the spines are shed, the nucleus undergoes repeated division filling the cytoplasm with daughter nuclei. These events are fairly uniform among the spinose species examined thus far. There appears to be only one nucleus in the planktonic, spinose foraminifera which gives rise to the daughter nuclei. No micronucleus has been conclusively demonstrated. The numerous daughter nuclei destined to be the nuclei of the swarmers become segregated in cytoplasmic islands within the intrashell cytoplasm by vacuolarization and lacunar formation (Figures 10, 12, 15 to 18). Thus interconnected masses of cytoplasm, containing one or more nuclei, are separated by meandering cisternae. In *G. ruber*, however, a stage preceding vacuolarization has been observed. Groups of daughter nuclei and some immediate surrounding cytoplasm, containing lipid droplets, mitochondria, Golgi, and endoplasmic reticulum, become

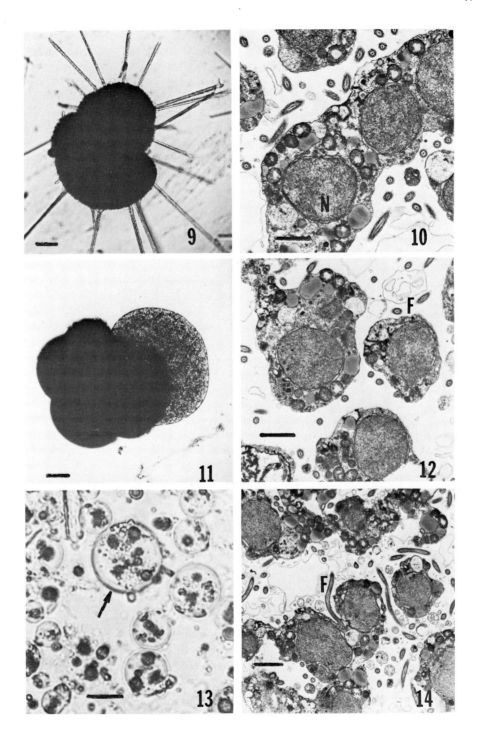

FIGURES 9 to 14. Major events during reproduction by swarmer release in *Hastigerina pelagica*. At an early stage (Figure 9, marker = 100 μm) the bubble capsule and spines are shed, and the intrashell cytoplasm becomes segregated into multinucleated (N) masses of cytoplasm (Figure 10, marker = 2 μm), the granular cytoplasm emerges from the shell aperture (Figure 11, marker = 100 μm) releasing numerous swarmers which have been separated from flagellated (F), nucleus-containing fragments of cytoplasm within the shell (Figure 12, marker = 2 μm). Thousands of spherical bodies (arrow, Figure 13, marker = 4 μm) containing a large vacuole with organic and/or calcitic debris are produced from a thin envelope of mother cell cytoplasm. Numerous flagellated swarmers (Figure 14, marker = 2 μm) are released through the shell aperture.

Table 2
DEVELOPMENTAL SEQUENCE OF SWARMER RELEASE IN *HASTIGERINA PELAGICA*[48]

Time before swarmer release (SR) and morphological changes

26 to 30 hr before SR — Lipid droplets become widely dispersed throughout the peripheral cytoplasm, no evidence of nuclear division.

22 hr before SR — Bubble capsule discarded, beginning with the distal bubbles; onset of chamber septal resorption and evidence of nuclear division. Specimens sink in culture vessel.

7 hr before SR — Spine resorption at the spine base, spine shedding commences, cytoplasm becomes multi-nucleated and segregated into inter-connected cytoplasmic islands.

5 hr before SR — Cytoplasmic bulge appears in aperture, cytoplasm becomes segregated into thin strands or clumps containing nuclei and bearing flagella.

0.5 hr before SR — Individual swarmers appear and "spherical bodies" are released.

SR — Cytoplasmic bulge bursts, releasing hundreds of thousands of biflagellated swarmers and 4000 to 6000 spherical bodies.

Table 3
DEVELOPMENTAL SEQUENCE OF SWARMER RELEASE IN *GLOBIGERINOIDES SACCULIFER*

Time before swarmer release (SR) and morphological changes

15 to 20 hr before SR — Spines are shortened, becoming stubs; symbionts are withdrawn into shell and some are lysed prior to expulsion of their remains. Lipid droplets become dispersed in the intrashell cytoplasm followed by onset of nuclear division.

12 hr before SR — Spine stubs are shed, extrashell cytoplasm forms a fine halo of fuzzy rhizopodia surrounding shell. Reproductive nuclei proliferate in intrashell cytoplasm, few intact symbionts remain.

8 hr before SR — Numerous nuclei fill the intrashell cytoplasm surrounded by vacuolated cytoplasm. Very few decrepid symbionts remain in large vacuoles. Cytoplasm appears milky white.

4 hr before SR — Clumps of multinucleated cytoplasm bearing flagella appear in the intrashell cytoplasm; extrashell cytoplasm appears granular with light microscope. Clumps of swarmers appear.

SR — A cytoplasmic bulge emerges and myriads of biflagellated swarmers are released. Shell becomes devoid of cytoplasm.

Note: These data are based solely on samples obtained offshore near the Bellairs Research Institute, St. James, Barbados, and cultured in the laboratory.

segregated from the remaining cytoplasm by a membrane boundary. As the number of nuclei increases, however, the boundaries are no longer detectable, and it is not clear whether this is a result of fusion of the bounded regions or as a result of incorporation of the surrounding cytoplasm into the membrane-bound multi-nucleated masses. In other spinose species, the nuclei appear to proliferate without specialized membranous barriers prior to the stage of vacuolarization and early segregation of swarmers.

Subsequent to the formation of nucleated cytoplasmic masses (Figure 17), flagella appear on the surface of the membranes lining the cisternae and project into the cisternal space. Finally, the nucleated cytoplasm is repeatedly divided into small, flagellated

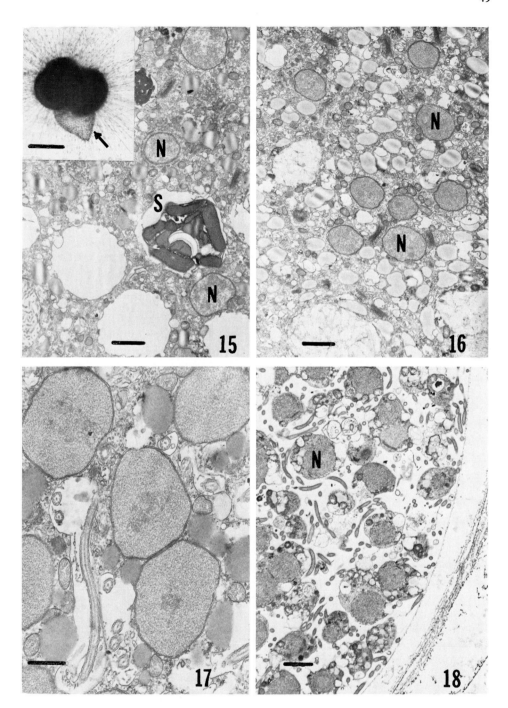

FIGURES 15 to 18. Cytoplasmic changes during swarmer production and release in *Globigerinoides sacculifer* with shell structure as shown in Figure 15 (inset), bearing a characteristic "sac-like" final chamber produced at maturity (marker = 300 μm). Approximately 12 hr before swarmer release (SR), vacuoles containing partially lysed symbionts (S) are observed in the intrashell cytoplasm and the cell nucleus begins to divide, producing daughter nuclei (N) dispersed throughout the cytoplasm (Figure 15, marker = 3 μm); the nuclei increase in number at 8 hr before SR when the cytoplasm becomes increasingly vacuolated and contains very few or no lysed symbionts (Figure 16, marker = 3 μm). At 4 hr before SR, clumps of nucleated cytoplasm bearing flagella occur within the shell (Figure 17, marker = 1.5 μm) and most symbionts or their lysed remains have been expelled. Ultimately myriads of the free swarmers, believed to be gametes, are released from the intrashell cytoplasm (Figure 18, marker = 3 μm) and escape through the aperture.

masses (ca. 3 to 5 μm) interconnected by thin cytoplasmic strands. In the last stages, the swarmers are released by separation from the main cytoplasmic mass and appear as biflagellated, uninucleate cells.

Swarmer production in *H. pelagica*, however, is further complicated by the appearance of spheroidal cytoplasmic bodies (ca. 10 to 30 μm diameter) containing a large central vacuole enclosed within a thin cytoplasmic envelope (Figure 13). The vacuole often contains organic and calcitic particles of unknown origin. The fine structure of the vacuole indicates it may be a residual digestive vacuole containing remnants of incompletely digested organic matter and perhaps excess hydrolytic enzymes. The nucleated cytoplasmic envelope possesses typical subcellular organelles observed in the incipient, swarmer-producing cytoplasm. The spheroidal bodies have not been observed to fuse with the swarmers or to differentiate further. They may be a means to eliminate residual digestive enzymes and waste products by enclosing them within a protective cytoplasmic envelope where they are separated from the delicate, developing swarmers. It is less likely that they are macrogametes given our present knowledge of their structure and activity. Swarmer release typically occurs in late afternoon for *H. pelagica* and is also frequently observed in *Globigerinella aequilateralis* and *Globigerinoides sacculifer* late in the day.

Globigerinoides sacculifer exhibits several unique features not observed in *H. pelagica*. The dinoflagellate symbionts (not present in *H. pelagica*) are withdrawn into the intrashell cytoplasm or proximal to the shell during the earliest stages of cytoplasmic contraction. Electron microscopic examination shows that many of the symbionts enclosed within vacuoles exhibit signs of lysis and are gradually decomposed before swarmer release. Other symbionts are expelled from the cytoplasm and appear in the light microscope as crenated yellow-brown bodies. It is not known whether the disintegration of the vacuole-bound symbionts is by autolysis or host digestion, or perhaps a combination of both. It is conceivable that some of the symbionts are digested by the host at the time of swarmer production to provide energy for the process and/or to make food reserves to be incorporated within the swarmer cytoplasm. Most well-nourished foraminifera possess copious amounts of lipid in their cytoplasm; therefore, it is not possible to detect additional food reserves that may come from symbiont lysis. Evidence of symbiont decay appears at the early stages of spine shortening and becomes more pronounced at the time of spine shedding. The fine structure evidence for cytoplasmic changes in *G. sacculifer* at periodic intervals during spine shortening, spine shedding and thereafter at hourly intervals is presented in Figures 15 to 18. This sequence provides an overview of the detailed events occurring at each time interval and permits correlation with the macroscopic events outlined in Table 3. The fine structure data clearly show that segregation of the cytoplasm into nucleated zones separated by lacunae occurs within a period of 3 to 4 hr after the spines are shed. Flagella begin to appear on the nucleated masses of cytoplasm between 4 and 6 hr after spine shedding and continue to increase in number and size until the eighth hour after spine loss. At this stage, much of the differentiated cytoplasm is separated into nucleated incipient swarmers attached to one another by thin cytoplasmic bridges. At the twelfth hour after spine loss, free swarmers are visible within the shell.

In all species of planktonic foraminifera examined, the swarmers are uninucleate, contain lipid reserve droplets, other typical cellular organelles observed in the mother cell cytoplasm, and possess two flagella inserted at one side of the cell.

Light microscopic evidence of swarmer production has also been observed in laboratory-cultured specimens of *Globigerinoides conglobatus, Orbulina universa, Globorotalia inflata, Pulleniatina obliquiloculata,* and *Globoquadrina dutertrei.*[30]

B. Radiolaria

1. Maturation

The complexity of cellular organization and degree of specialization observed in solitary cells of radiolaria becomes considerably elaborated when multicellular associations are formed. Colonial radiolaria possess hundreds to thousands of central capsules bound together in a translucent gelatinous envelope penetrated by a fine network of rhizopodial strands interconnecting the radiolarian cells and enclosing the algal symbionts. A fine halo of peripheral rhizopodia projects from the gelatinous envelope to enhance prey capture or facilitate exchange of nutrients and wastes with the surrounding environment. In *Collozoum caudatum*[44] some of the rhizopodia become further specialized as internal transporting pathways by forming a cable-like strand running axially through the filiform colony. This strand may facilitate food dispersal within the colony and also carry waste products to the tips of the colony for elimination.

During early stages of reproduction in radiolaria, the specialized cytoplasmic regions are transformed in a variety of ways among different species to cause sinking and facilitate swarmer release. There are sufficient differences in maturation and reproduction between solitary and colonial radiolaria to warrant separate consideration.

Little is known about early stages of solitary radiolarian development, although some research has begun in our laboratory. The diversity of species makes it difficult to generalize findings from examination of a limited number of specimens. Some solitary radiolaria possess siliceous skeletons of remarkable complexity and architectural intricacy[25,35,36] that suggest considerable diversity in skeletal ontogeny. In addition to species with spherical or polyhedral lattice skeletons, some radiolaria possess a spongiose skeleton composed of a fine mesh of sponge-like silica deposited on a central sphere or on a polyhedral framework. The central capsular wall, enclosing the nucleus and its surrounding cytoplasm, appears to increase in diameter during ontogenesis and sometimes becomes indented or deformed at points of contact with the skeleton as the wall expands. The ontogeny and maturation of colonial radiolaria has been reported for only a few species, and the data have been largely assembled from observations of developmental stages gathered from the open ocean supplemented by laboratory investigations of transformation of stages in short-term cultures. The earliest stage of development in *C. caudatum*[44] (a non-skeletal species) appears to be a solitary central capsule surrounded by a thick gelatinous envelope containing the symbionts and a rhizopodial network. The central capsule elongates and divides into several prolate central capsules enclosed within a cylindrical gelatinous envelope that is secreted as the radiolarian cells in the colony increase in number. The prolate central capsules eventually give rise to numerous spheroidal central capsules and the colony assumes a characteristic "sausage" shape with a string of central capsules gathered as a line along the central axis of the colony. Skeletal-bearing species, e.g., collosphaerid species may begin development as a small cluster of several central capsules derived from a single-celled form.[40] The earliest identifiable stages (containing a score or more cells) are largely skeletonless, except for a few peripheral cells that possess silica within the cytokalymma bearing a characteristic skeletal shape, thereby confirming their species identity. The skeletons are either deposited by a process of bar growth producing first a hollow lattice sphere that is progressively subdivided into smaller polygonal pores as in *C. huxleyii* or by rim growth where the pores are thickened at the rim and become smaller with clearly circular outlines as in *Siphonosphaera tubulosa*.[40] Neither the rate of skeletal maturation nor the quantity of silica deposited per unit of time is known for most radiolaria. Longevity in laboratory culture has been determined for several skeletal and non-skeletal species of solitary radiolaria.[35] *Spongodrymus* sp. live for periods up to 21 days while *Thalassicolla nucleata* lives for periods up to 30 days depending on availability of light and food, and their maturity at the time of collection. Since all of the specimens cultured in the laboratory were collected from the ocean at near maturity, they undoubtedly have even longer lifespans than those reported from current laboratory investigations. These studies, however, have shown that radiolaria can live far longer than had been presumed heretofore and that the maturation phase of their life cycle is on the order of several weeks rather than days. Reproduction is by swarmer release as in planktonic foraminifera; however, swarmer morphology and cytoplasmic fine structure are very different. The general events during swarmer development and the fine structure of reproducing organisms have been determined for some solitary and colonial species.

2. Reproduction

a. Solitary Species

The general sequence of events during swarmer production in representative solitary, skeleton-bearing species *Spongodrymus* sp. and a non-skeletal species *Thalassicolla* sp. will be presented. The earliest detectable evidence of reproduction in laboratory-cultured specimens is refusal to accept prey as is also observed in spinose planktonic foraminifera. Within a period of several hours, the extracapsular rhizopodial network is shed or withdrawn and the radiolarian loses buoyancy, settling to the bottom of the culture vessel, also, as in planktonic foraminifera. The extracapsulum is shed very rapidly in the non-skeletal species *T. nucleata*. A normal-appearing, floating specimen will suddenly cast off its extracapsulum. The naked, black-pigmented central capsule settles to the bottom of the culture dish within minutes. The extracapsulum disintegrates upon release and cannot be detected by visual inspection. Skeleton-bearing species sometimes settle more slowly and possess a thin layer of extracapsulum which is gradually withdrawn over several hours.

The settled central capsule of many radiolarian species appears distinctly opalescent white and any surface pigment, which may vary from an irridescent violet to purplish brown or red, is lost. The central capsule in *T. nucleata* becomes pearly white and glossy in appearance. Within a period of twelve hours after shedding the extracapsulum in *T. nucleata*, a milky white bulge of cytoplasm protrudes through a fissure in the capsular wall and may produce a finely filamentous mass of cytoplasmic threads containing the incipient swarmers.[26] The flagellated swarmers emerge from the bulge and swim away from the mother cell with a slight undulating or gyrating motion. In spongiose skeletal species, the swarmers are released from the central capsule and emerge from the spongiose skeleton which obscures the central capsule cytoplasm. Each of the myriads of swarmers appears decidedly angular in profile due to a rectangular-prismatic crystal enclosed within a vacuole. None of the swarmers have been observed to fuse and therefore, we do not know whether the sluggishly swimming swarmers are asexual spores or gametes.

The fine structure evidence for swarmer production and release is described in conjunction with data for colonial species in a section to follow.

b. Colonial Species

Early signs of swarmer production in colonial radiolaria include: (1) thinning of the translucent gelatinous envelope which becomes increasingly soft and jelly-like, (2) contraction of the rhizopodial system resulting in withdrawal of the peripheral threadlike filopodia, shrinkage of the colony and the aggregation of central capsules into compact clumps, and (3) gradual disintegration of the gelatinous envelope. During early stages of shrinkage, the colony loses buoyancy and gradually sinks to the bottom of the culture vessel over a period of several hours. The progressive dissolution of the gelatinous matrix surrounding the colony and the rhizopodial contraction causes the central capsules to become aggregated, thus further increasing the density of the reproducing colony. When the colony contains numerous central capsules, they are gathered by rhizopodial contraction into small clumps of about 20 or more cells. The clumps break apart and separate from one another as the gelatin dissolves and the rhizopodial network connecting the cells disintegrates. These events have been observed most closely in *Collosphaera globularis* and *Sphaerozoum punctatum*.[45] During rhizopodial contraction in *S. punctatum*, some of the numerous triradiate symmetrical spines (Figure 4) become concentrated around the central capsules and remain there until the latest stages of rhizopodial disintegration. The clumps of central capsules are interconnected by thin cytoplasmic strands and remain in contact until the onset of swarmer release. In many cases, the reproducing central capsules do not develop in phase, and one or more begin releasing swarmers earlier than the others. The development of the swarmers within the central capsule can be monitored by light optics. The refractile rectangular prismatic crystal in each swarmer is visible through the thin capsular wall, and their gradual growth is clear evidence of swarmer development and maturation. When the swarmers are mature, they become free and motile while enclosed within the capsular wall. Within minutes after the swarmers become active, the capsular wall ruptures and the swarmers are released. A summary of events during swarmer production in solitary and colonial radiolaria is presented in Table 4.

c. Fine Structure

The major fine structure events during swarmer production and release are sufficiently comparable in solitary and colonial species that they can be described together. In many solitary and colonial radiolaria, the intracapsular cytoplasm is segregated into interconnected lobes either arranged radially (Figure 8) or intertwined with narrow cisternae between them. Solitary species usually possess a single nucleus during maturation, but become multinucleate at reproductive maturity. Colonial species (e.g., *Sphaerozoum punctatum*, *Collosphaera globularis*, and *Collozoum inerme*), however, are frequently multinucleate well in advance of swarmer production. The nuclei proliferate by repeated division and gradually increase in number as the cell matures. Dividing nuclei have been observed in *S. punctatum*.[45] The nuclear envelope is persistent during karyokinesis. The spindle fibers, composed of microtubules, are anchored on the inner membrane of the nuclear envelope at a thickened site, and intersect with chromatin in the nucleoplasm. A nucleus in late stage of division presents a distinctive dumbbell profile (Figure 19) produced by a constriction ring of microfilaments encircling the nucleus in the division plane. No centriole has been observed, but an osmiophilic body lying outside the nucleus near the spindle-fiber attachment site may be a microtubule organizing center. Dividing nuclei in planktonic foraminifera also possess a persistent nuclear membrane and contain intranuclear spindle fibers. The chromatin, however, is seldom condensed into cord-like masses as observed in some radiolaria.

In uninucleate solitary species, such as *Thalassicolla nucleata*, the centrally located nucleus begins dividing at the early stages of reproduction before the extracapsulum is shed. The closely packed lobes of intracapsular cytoplasm become occupied by daughter nuclei and subsequently separate into multi-nucleated masses interconnected by cytoplasmic strands. In colonial radiolaria, the multinucleated lobes become more widely separated and gradually separate into multinucleated masses. Early stages of swarmer production in *S. punctatum* and *T. nucleata* are shown in Figures 20 and 21 respectively.[26,45] At these very early stages, flagella may not be present, but the vacuoles (V, Figure 20) destined to contain the crystalline inclusions are already forming. Cytoplasmic differentiation ultimately produces masses of interconnected cytoplasmic lobes bearing flagella projecting into the intervening cisternae. This stage is remarkably

Table 4

A GENERALIZED DEVELOPMENTAL SEQUENCE OF SWARMER RELEASE IN RADIOLARIA

Time before swarmer release (SR) and morphological changes

12 to 20 hr before SR — Feeding rhizopodia are withdrawn or prey is rejected; intracapsular cytoplasm becomes filled with daughter nuclei commencing earlier in many colonial species compared to solitary species; extracapsulum is shed abruptly in *Thalassicolla* or over several hours in some skeletal-bearing species; colonial forms begin contraction of rhizopodia and shrinking of gelatinous envelope. Radiolarian loses buoyancy and settles in water column.

8 to 10 hr before SR — Intracapsular cytoplasm becomes increasingly vacuolated, and cisternae appear between masses of cytoplasm containing nuclei often with cord-like chromatin strands. Colonial species fragment into small clumps of central capsules possessing a milky white appearance.

4 to 6 hr before SR — Incipient swarmers and clumps of partially differentiated cytoplasm within central capsules contain developing crystals of strontium sulfate which enlarge as the swarmers mature.

1 to 2 hr before SR — Flagellated masses of cytoplasm and some free swarmers appear within central capsules.

SR — Central capsule wall ruptures and a milky bulge of cytoplasm emerges releasing swarmers from its surface, or the already motile swarmers burst forth from the ruptured capsule.

similar to parallel stages of development in planktonic foraminifera (Figures 14 and 17). However, the presence of the crystal-containing vacuoles, and the characteristic nuclei with cord-like masses of chromatin in some radiolarian species clearly distinguish radiolarian cytoplasm from planktonic foraminiferal cytoplasm. The crystal inclusions in the cytoplasmic masses of incipient swarmers are brittle and fracture during sectioning leaving a characteristic lenticular or parallelogram-shaped hole in the Epon section (X, Figure 22). When the swarmers are mature, the capsular wall (limiting peripheral assembly of plasma membranes, CW in Figure 20) ruptures along sites of minute fissures located in the wall between the fusules. The dislocated segments permit release of the motile swarmers (Figure 22), possessing a prominent nucleus, mitochondria, lipid droplets, Golgi bodies, other typical cytoplasmic organelles, and the large crystal enclosed within a vacuole. X-ray dispersion analysis with the scanning electron microscope shows the presence of strontium in the crystals, but not silica.[46,26] The symmetry of the crystal examined by light microscopy indicates that it is strontium sulfate.[3] We presently do not know the function of this crystal which is sufficiently large to occupy close to 1/3 of the volume of the cytoplasm in a swarmer. It is interesting to note, however, that Acantharia, a closely related group of protozoa, also possess strontium sulfate deposits in their skeletons which may indicate a common ancestry with the radiolaria.[36]

IV. ENVIRONMENTAL AND PHYSIOLOGICAL FACTORS AFFECTING ONSET OF REPRODUCTION

There appears to be a threshold level of physiological development that must be reached in planktonic foraminifera and radiolaria for reproduction to occur, as is also observed in higher organisms. Juvenile specimens seldom form swarmers in laboratory culture and unless conditions are favorable for growth, they die without reaching reproductive maturity. For example, specimens of *Globigerinoides sacculifer* very seldom produce swarmers in laboratory culture if they cease to grow beyond 300 μm.[47]

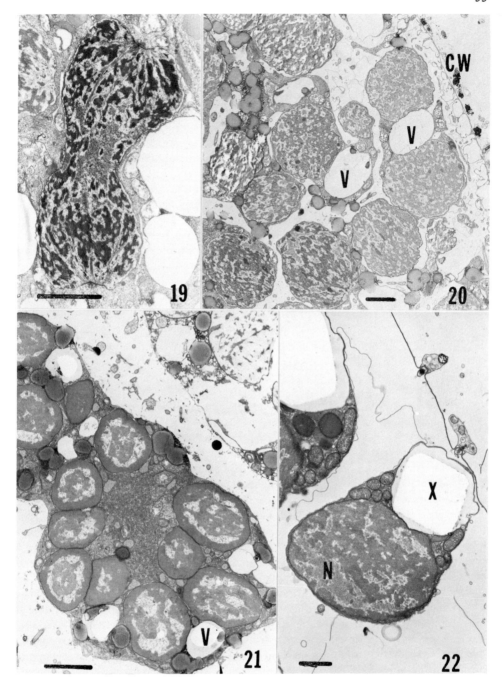

FIGURE 19. A dividing nucleus in the colonial radiolarian *Sphaerozoum punctatum* displays a dumbbell-shaped profile produced by constriction of the persistent nuclear envelope in the median zone. Microtubules attached to the nuclear membrane intersect with masses of condensed chromatin. Marker = 4 μm. (From Anderson, O. R., *Mar. Micropaleontol.*, 1, 287, 1976. With permission.)

FIGURES 20-21. Early stages of swarmer production in *Sphaerozoum punctatum* (Figure 20) and *Thallisicolla nucleata* (Figure 21) exhibit masses of multinucleated intracapsular cytoplasm bearing vacuoles (V) destined to contain crystalline inclusions. The capsular wall (CW) eventually ruptures, releasing the flagellated swarmers. Markers = 4 μm and 2 μm.

FIGURE 22. A free crystal-bearing swarmer from *Sphaerozoum punctatum* possesses a prominent nucleus (N) with cord-like masses of chromatin, and a large vacuole containing the profile (X) of the crystalline inclusion that has fallen out during sectioning. Marker = 2 μm. (From Anderson, O. R., *Mar. Micropaleontol.*, 2, 251, 1977. With permission.)

Radiolaria also exhibit a maturational threshold for reproduction. Among hundreds of *Thalassicolla nucleata* specimens observed in our laboratory, clear evidence of swarmer production was observed only in specimens with mature central capsules (Ca. 0.7 to 0.9 mm diameter). In general, we have observed that the larger the diameter of *T. nucleata* or *Spongodrymus* sp. at the time of field collection, the earlier the specimen produces swarmers in the laboratory. Mature specimens (capsules = 0.8 mm or larger) will usually release swarmers within 24 hr after collection, whereas smaller specimens may live for three or four days before releasing swarmers or even up to several weeks if their size is less than 0.7 mm diameter and adequate light and food are available for nourishment. We do not know what physiological factors trigger swarmer production when a radiolarian reaches reproductive maturity. Possible variables include cell size and concomitant accumulation of sufficient food reserves, a genetic-developmental threshold, cytoplasmic metabolic rate, or production of metabolic products that regulate the nuclear cycle by feedback effects.

There is clear evidence in spinose planktonic foraminifera that a combination of environmental and physiological variables influence onset of swarmer production once the organism has reached reproductive maturity. These include: (1) a lunar periodic stimulus in *Hastigerina pelagica*, (2) light-stimulated growth and triggering of reproduction mediated by symbiont photoreception in *G. sacculifer*, and (3) rapid growth due to frequent feeding in *G. sacculifer*. Experimental evidence for each of these regulatory factors in planktonic foraminifera is discussed, and pertinent comparative data from research with radiolaria are also presented.

A. Lunar Periodicity

Combined evidence from field observations and analysis of reproducing specimens in laboratory culture show that *H. pelagica* produces swarmers on a regular cycle coinciding with the period of a full moon. Within 3 to 7 days past a full moon, a majority of mature specimens of *H. pelagica* observed in the natural environment, and those that were recently collected and maintained in laboratory culture, produced swarmers.[48,49] Immediately following these events, few mature specimens were found in the surface water of the ocean. The swarmers are apparently released at greater depths. The mature *H. pelagica*, devoid of a bubble capsule, sinks in the water column and releases swarmers after settling for 12 or more hr. Juvenile specimens gradually begin reappearing in the surface water within one week and become increasingly larger as time progresses toward the next full moon. In 717 instances of swarmer release by laboratory cultured specimens collected near Bermuda or offshore at Barbados, 85.4% of the individuals produced swarmers within 3 to 7 days after a full moon. These observations span a period of three years (July 1975 to June 1978). The number of individual *H. pelagica* that produced swarmers within the period of 10 days after a full moon is presented in Table 5. The distribution approximates a normal curve with a mean of 5 days past a full moon for swarmer release.[49] These data gathered each month during spring and summer when *H. pelagica* is most abundant near Bermuda, and during winter in Barbados, show a consistent lunar cycle for *H. pelagica* reproduction. Occasional specimens in the laboratory fail to reproduce at the first occurrence of a full moon but reproduce 29 days later at the next full moon. This suggests that the generation time may also be a multiple of 29 days for some individuals. Indeed, a few mature *H. pelagica* are always found remaining in the ocean surface water after the majority of the population have reproduced. Although these combined laboratory and field observational data clearly point to a lunar reproductive cycle in *H. pelagica*, the physiological mechanism for the triggering of swarmer production is not fully understood. Further research is needed on the source of the rhythm for the physiological clock and possibly on photo-sensory receptive processes or gravitational field percep-

Table 5
NUMBER OF *HASTIGERINA PELAGICA* SPECIMENS
RELEASING SWARMERS DURING A TEN DAY PERIOD
FOLLOWING A FULL MOON (DAY 0)[49]

Days	0	1	2	3	4	5	6	7	8	9	10
Specimens	1	20	16	130	115	165	135	60	15	42	2

tion that may mediate the periodic, lunar reproductive response. At present we have not found evidence of lunar periodic reproduction in other planktonic foraminifera or in the radiolaria we have examined (e.g., *Spongodrymus* sp., *Hexastylus* sp., *Thalassicolla nucleata, T. melacapsa* or several colonial species). However, much additional data will be required before a more firm conclusion can be reached for radiolaria.

B. Light-mediated Onset of Reproduction

Globigerinoides sacculifer has been one of the more supple and robust planktonic foraminifera in laboratory culture. Consequently, we have much more information about its vital processes than for other species. In the section on cellular specialization, it was mentioned that *G. sacculifer* establishes a close structural association with its dinoflagellate symbionts mediated in part by the intimate cytoplasmic union made possible by the transitional organization of the host cytoplasm. The absence of barriers between large regions of the cytoplasm enhances the possibility of symbionts being carried into the central cytoplasm of the host through the shell aperture, thus establishing close contact with the intrashell cytoplasm. There is clear evidence that this intimate structural association of symbiont and host is further reinforced by a close physiological interaction. Symbionts may provide organic nourishment to the host by transferring photosynthetic products during enclosure within the intrashell cytoplasm. Osmiophilic substances resembling those within the periplast membranes surrounding the symbiont are present in host vacuoles shortly after the symbionts are withdrawn into the shell at the end of the day.[24]

Experimental studies on the role of light intensity in maturation and reproduction of *G. sacculifer* provide additional evidence of a close relationship between symbiont and host.[47,50,51] When *G. sacculifer* (150 to 500 μm size) is cultured in total darkness, they fail to grow as large as equivalent specimens grown in the light (Table 6).[50] Moreover, the intensity of the light is directly correlated with growth as measured by increase in the number of chambers. The illumination was varied from a moderate level of 20 to 50 μE/m^2/sec to a higher level of 400 to 500 μE/m^2/sec. These data suggest that either the host senses the light and is directly influenced by it, or that the symbionts are in part mediating growth through primary productivity and stimulation of the host, perhaps by translocation of photosynthates to the host.

To test whether the symbionts are a major contributing factor in light-stimulated growth of *G. sacculifer*, young specimens were treated with an inhibitor of photosynthesis, 3-(3,4 dichlorophenyl)-1, 1-dimethyl urea (Diuron) for 72 hr to eliminate the symbionts. Electron microscopic examination of the specimens confirmed that the symbionts were disabled and lysed due to inhibition of photosynthesis.[31] Diuron treatment caused a gradual depletion in carbohydrate reserves and eventual death of the symbionts. No major influence on host cytoplasm was detected. As with dark cultured specimens, Diuron-induced aposymbiotic *G. sacculifer* also failed to grow appreciably. Moreover, feeding the aposymbiotic foraminifera with *Artemia* nauplii, a normally productive food source, also failed to normalize growth. When, however, the aposymbiotic *G. sacculifer* were reinfected with symbionts from a donor *G. sacculifer*, growth resumed and the specimen eventually achieved a size comparable to natural symbiont-

Table 6

EFFECTS OF VARIATION IN LIGHT INTENSITY ON GROWTH AND REPRODUCTION IN *GLOBIGERINOIDES SACCULIFER* [50]

	Mean shell size (µm)		Mean longevity (days)	Frequency of sac chambers (%)	Total chambers formed	Frequency of swarmer production (%)
	Initial	Final				
High Intensity	323	791	10	73	188	94
400—500 µE/m²/sec	(102)	(76)	(3)			
Moderate Intensity	324	707	7	76	157	95
20—50 µE/m²/sec	(102)	(131)	(3)			
Continuous darkness	325	474	4	41	86	92
	(103)	(85)	(2)			

Note: Specimens were all fed one *Artemia* nauplius daily. Number in parenthesis is standard deviation. Mean longevity is the mean number of days the specimen lived in culture; the final, sac-like chambers are characteristic of reproductive maturation; frequency of swarmer production is percent of specimens producing swarmers in each treatment. Light cycle is 12 hr light/12 hr dark. There were 63 specimens per treatment.

bearing controls.[51] These data clearly indicate that the dinoflagellate symbionts in *G. sacculifer* serve a critical enabling function in host growth and maturation. Moreover, the stimulatory influence appears to be more than a general nutritional role, since supplemental feeding with *Artemia* as food in the absence of symbionts failed to restore normal growth; whereas a comparable feeding regime in symbiont-bearing hosts promoted regular shell growth. To determine if the limiting factor for growth was merely lack of algal food, dark-grown *G. sacculifer* were fed a mixture of free-living algae including *Amphidinium carterae*, *Coccolithus huxleyi*, and *Isochrysis galbana*. This algal diet failed to reverse growth inhibition and eventual early swarmer production. These data raise the intriguing question of whether the symbionts serve a regulatory function in a much more significant way, perhaps through secretion of messenger or hormonal-like molecules. Experimental evidence is meager at the present time but some interesting insight has been obtained from analysis of mature *G. sacculifer* and their onset of swarmer production in relation to presence or absence of light. If symbiont-bearing mature *G. sacculifer* (shell size, >200 µm) are placed in total darkness, they commence swarmer production within 3 to 5 days. Equivalent control samples maintained in the light can continue to mature for as long as 10 to 30 days depending on the feeding frequency as explained in the next section. It is clear from these data that absence of light induces onset of swarmer production in mature *G. sacculifer* that otherwise would live for a longer time when light is present. To determine whether the symbionts provide a mediating factor in the early onset of swarmer production, mature, symbiont-bearing *G. sacculifer* were treated with Diuron to render them aposymbiotic. The "bleached" specimens, maintained in standard light conditions and with controlled nutrition as in the dark-treated specimens, also commenced swarmer production within 3 to 5 days after symbiont elimination. This early commencement of reproduction can be prevented, however, by transferring living symbionts from a donor *G. sacculifer* to the aposymbiotic specimens. These data provide additional evidence that the symbionts serve a significant photo-receptor function in mediating light effects on growth and reproduction in *G. sacculifer*.[51]

By contrast, the role of symbionts in radiolarian reproduction is largely unexplored. Preliminary evidence with *T. nucleata* show that dark treatment does not induce early onset of swarmer production and that specimens fed with *Artemia* while in total dark-

Table 7
EFFECT OF FEEDING FREQUENCY ON GROWTH AND
REPRODUCTION IN *GLOBIGERINOIDES SACCULIFER* [47]

Feeding frequency	Mean shell size (μm)		Mean longevity (days)	Frequency of sac chambers (%)	Total chambers formed	Frequency of swarmer production (%)
	Initial	Final				
Daily	323	791	10	73	188	94
	(102)	(76)	(3)			
Every 3 days	326	705	14	78	161	92
	(102)	(90)	(6)			
Every 7 days	324	584	23	32	107	84
	(100)	(96)	(15)			
Unfed	323	289	31	2	102	32
	(103)	(176)	(18)			

Note: Culture conditions are at high light intensity (400 to 500 μE/m²/sec).

Specimens were all fed one *Artemia* nauplius daily. Number in parenthesis is standard deviation. Mean longevity is the mean number of days the specimen lived in culture; the final, sac-like chambers are characteristic of reproductive maturation; frequency of swarmer production is percent of specimens producing swarmers in each treatment. Light cycle is 12 hr light/12 hr dark. There were 63 specimens per treatment.

ness live almost as long as control specimens kept in a 12 hr light/dark treatment and fed on the same schedule. As might be expected, unfed symbiont-bearing specimens maintained in the light lived longer than equivalent size controls kept in continuous darkness. These data suggest that, as in *G. sacculifer*, the symbionts provide nourishment to the host by supplying photosynthates. Laboratory experiments with colonial species confirm that ^{14}C-labelled photosynthates are transferred to the host,[34] and more recently that solitary species (*T. nucleata*) are also recipients of symbiont-derived organic nourishment. In these experiments, illuminated, unfed, symbiont-bearing *T. nucleata* lived for periods up to 4 or 6 weeks, suggesting that the symbionts provide a significant part of the basic nourishment of the host.

C. Effects of Feeding Frequency on Reproduction

The frequency of feeding (i.e., once each day, every 2 days, every 3 days, etc.) using *Artemia* nauplii as prey has a profound influence on maturation and reproduction in *G. sacculifer*.[47] Shell growth as measured by increase in shell size and number of chambers added is directly correlated with feeding frequency (Table 7). For example, the mean final shell size attained by *G. sacculifer* was 791 μm for a daily feeding regime, 705 μm for a 3-day feeding frequency, and 584 μm when fed every 7 days. The mean initial size of the specimens in each of the 3 feeding regimes was 324 μm. Daily feeding produces a clear enhancement of growth, whereas unfed individuals ceased growing and resorbed or shed chambers thus yielding a diminished mean final size of 289 μm.

Although growth is enhanced by more frequent feeding, the rapid maturation also results in a shorter life period (Table 7). Frequent feeding results in an earlier reproductive maturity as evidenced by formation of the characteristic sac-like chamber and release of swarmers. Less frequent feeding results in prolonged, and less rapid, growth that causes delayed reproductive maturation and hence a longer life period in culture. As an illustrative comparison, specimens of *G. sacculifer* that were fed every day (final shell = 791 μm) lived in culture for a mean of 10 days after the inception of the experiment; whereas individuals fed every 7 days (final shell = 584 μm) lived on the average

Table 8
MEAN SURVIVAL TIME AND
VITALITY SCORE FOR *HASTIGERINA
PELAGICA* AS A FUNCTION OF
FEEDING INTERVAL[29]

Feeding interval	Mean survival (days)	Vitality score (mean days floating)
Daily	23.2	17.7
Six days	26.8	21.7
Twelve days	17.4	12.9
Starved	16.4	13.7

Note: N = 30 specimens per group.

23 days in culture. Under oligotrophic conditions, therefore, *G. sacculifer* tends to grow more slowly, but lives longer in comparison to those in eutrophic environments. These data indicate that the duration of the maturational phase in the growth cycle of *G. sacculifer* is partially regulated by the availability of food. An eutrophic environment results in shorter maturational phases and more frequent turnover of the reproductive cycle than in oligotrophic environments. Less frequent feeding extends the maturational phase in each generation and decreases the number of reproductive cycles per unit of time.

The effects of feeding frequency on maturation and reproduction in *H. pelagica* are not so pronounced as in *G. sacculifer*. Moreover, the longevity in laboratory culture is directly related to the frequency of feeding as shown in Table 8.[29] The mean longevity for daily feeding with one *Artemia* nauplius was 23.2 days, whereas feeding every 6 or 12 days yielded a mean longevity of 26.8 and 17.4 days, respectively. It is clear also from the data in Table 8, that the health of the specimens as assessed by buoyancy was considerably better for specimens fed every day or every six days compared to specimens fed every twelve days or not fed at all.

It is not surprising that a frequent feeding interval enhances vitality and longevity in *H. pelagica*, but does not influence the onset of reproduction. As explained in the previous section, the reproductive cycle in *H. pelagica* is regulated by a lunar periodicity. This is apparently a very conservative mechanism which is not perturbed appreciably by environmental variables such as feeding frequency. If the organisms receive sufficient food to reach maturity, the lunar periodic cycle is the major determining factor in onset of swarmer production. In the interim, however, the vitality and longevity of *H. pelagica* depends on the frequency of feeding and perhaps on the quality of the food. More research is needed on the effects of food quality (including contributions of variety in natural zooplankton prey and perhaps phytoplankton prey) on the physiology of *H. pelagica*, as we have largely used *Artemia* nauplii in our laboratory experiments.

Among the radiolarian species examined thus far, the quality of the food offered (i.e. carnivorous, omnivorous, or herbivorous diets) appears to be equally or more significant than feeding frequency in swarmer production. *Spongodrymus* sp. (a large, spongiose skeletal species) lives longer in laboratory culture and exhibits more frequent evidence of swarmer production when offered an omnivorous diet as compared to a carnivorous or herbivorous diet. Preliminary data obtained in collaboration with Dr. N. R. Swanberg shows that *Spongodrymus* species offered a mixture of *Artemia* nauplii and algae (*Coccolithus huxleyi, Isochrysis galbana,* and *Amphidinium carterae*) survived for a mean of 18 days in culture, whereas the mean longevity for specimens fed carnivorous and herbivorous diets was respectively 14 and 9 days. The mean survival of *T. nucleata* in laboratory culture is not significantly influenced by the kind of food offered within the limitations of the crustacean and algal prey examined (*Artemia* nauplii and/or algal prey). This may be attributed to the significant trophic role served by the symbionts. Given the numerous species of solitary and colonial radiolaria (perhaps over one thousand valid species), considerable additional research is needed before conclusive statements can be made on the effects of nutrition on maturation, reproduction, and/or survival in radiolaria.

V. ADAPTIVE SIGNIFICANCE OF REGULATORY MECHANISMS IN SWARMER PRODUCTION

Given our limited understanding of the life cycles of planktonic foraminifera and radiolaria, it is not possible to make firm conclusions about the adaptive significance of environmental and physiological factors influencing reproduction as cited in the previous section. Certain theoretical principles emerge, however, when these regulatory events are examined in the context of the ecology and physiology of planktonic foraminifera and radiolaria.

One of the most striking common features of swarmer production in planktonic foraminifera and radiolaria is the loss of buoyancy and sinking of the reproducing organism in the water column. The mechanism of reducing buoyancy and settling differs markedly between planktonic foraminifera possessing a transitional specialized organization and the radiolaria with zonal specialization. During early stages of reproduction, the planktonic foraminifera gradually withdraw their rhizopodia and contract their extrashell cytoplasm promoting settling within the water column. Consequently, many of the lysosomes, residual digestive vacuoles and symbionts are mixed into the intrashell cytoplasm where swarmer production occurs. This is fully consistent with a transitional specialization pattern where the cytoplasm possesses few barriers to separate digestive function from other specialized processes. The contraction of the extrashell cytoplasm near the shell surface and into the intrashell space results in mixing of the symbionts and extrashell organelles with the inner shell cytoplasm. Subsequently, during swarmer production and release, these potentially destructive lytic vacuoles charged with wastes must be eliminated. In *H. pelagica*, the potentially destructive lysosomes and residual vacuoles are enclosed in the specialized spherical bodies and jettisoned. The symbionts in *G. sacculifer* are withdrawn into the cytoplasm and become degraded or expelled in a decrepid state before swarmer production is completed. Small residual digestive vacuoles remain enclosed in protective cytoplasmic envelopes in the intrashell cytoplasm. Radiolaria such as *T. nucleata*, with zonal specialization are able to segregate potentially destructive or interfering bodies from the developing swarmers by casting off the extracapsulum entirely and settling in the water column with the protective capsular wall enclosing the delicate developing swarmers. Hence, swarmer production can proceed without impedance from potentially autolytic digestive vacuoles. This is further evidence of the advanced state of specialization in radiolaria where the intracapsulum is largely converted to a reproductive center at the time of swarmer production.

It is not possible to determine how deep the reproducing organisms sink before swarmers are released as both the rate of sinking and time of swarmer maturation after loss of buoyancy may vary considerably from species to species. Based on preliminary laboratory investigations of sinking rates in planktonic foraminifera, it is possible that the reproductive organism descends to depths of 50 to 200 meters or more before swarmers are released.[42] We do not know what effect a transitory or more permanent physical boundary such as a thermocline may have on settling rates or indeed, if the reproductive organisms settle into or pass through the thermocline.[52] It is clear, however, that swarmers may be released at somewhat greater depths than those inhabited by the mature reproductive organism.[42] Various adaptations have evolved to enhance sinking of the reproducing organism while enabling swarmer release. The intrashell septa and spines of *H. pelagica* may be resorbed during swarmer maturation to aid expulsion of the mature swarmers.[48] *G. sacculifer* and other species, however, may thicken its shell by adding additional layers of calcite, thus increasing its density.[42] Among the radiolaria, the strontium sulfate inclusions in the developing and mature swarmers undoubtedly increases the density of the reproducing organism and also contributes to further sinking of the swarmers when they are released.[36,45] If the reproducing organism settles into the biologically-enriched zone of the thermocline where physico-chemical analyses indicate a chlorophyll maximum,[52] the offspring resulting from the swarmers would have ample sources of phytoplankton and microzooplankton prey. The settling phenomenon, moreover, may carry reproducing organisms into deeper water, where there are fewer predators than in the surface water.

There is a fair amount of evidence however, to show that descent into deep water may not be a necessary condition for development of offspring in *Globigerinoides ruber*. Mature specimens, introduced into water-permeable bags anchored in open ocean locations at about 3 meters depth, produced viable offspring as evidenced by the appearance of juveniles and their subsequent maturation. Hence, physical factors such as increased pressure, decreased temperature, and/or diminished light intensity may not be necessary in promoting second generation development in some species of planktonic foraminifera. We do not know at what point the swarmers, which lack symbionts, and/or their progeny, capture new symbionts. Some foraminifera and radiolaria consistently possess symbionts when collected in the natural environment, e.g., *Globigerinoides sacculifer*, *G. ruber*, and *Orbulina universa* among planktonic foraminifera to *Spongodrymus* sp., *Collozoum inerme*, *Collosphaera globularis* and many others among the radiolaria. The vertical descent in the water column may provide enhanced opportunity to capture symbionts by bringing the offspring into regions of enriched symbiont densities or where the potentially symbiotic algae are in the proper developmental stage to be captured by the young host.

The lunar periodic reproductive cycle in *H. pelagica* is an intellectually tantalizing phenomenon. Aside from the myriad questions about the phylogenetic origin, physiological mechanism, and ontogenetic development of this cycle, there is clear evidence that the phenomenon synchronizes swarmer release among a substantial number of mature individuals within a population at each 29-day period. If the swarmers are gametes as suggested by morphological and behavioral evidence, their simultaneous release among numerous individuals would increase the probability of syngamy by increasing their density per unit volume. Synchrony may also promote cross-fertilization by favoring mixing of gametes from many different mother cells. There are always a few individuals in the population that do not reproduce in synchrony with the majority of reproductive organisms. They may be out of phase by one or two days or persist beyond the lunar period until the next full moon. These residual organisms may provide a mechanism to increase the probability of population continuity in the event of a natural disaster or unfavorable growth environment that could deplete a large part of a new generation. Hence the advantages of synchronized gamete release for syngamy and cross-fertilization is also fraught with the disadvantage of massive loss of offspring in a coincident disaster. The residual group of non-synchronized, mature organisms can partially compensate for such an untoward event and thereby enhance continuity of the gene pool. It is not surprising that among the other highly specialized traits exhibited by *H. pelagica*, it also possesses a remarkably sophisticated mechanism for promoting synchrony in reproduction.

The trophic-regulatory influence in *G. sacculifer* is further evidence of environmental synchronizing events. When food is abundant, the foraminifera mature rapidly and reproduce in closer synchrony than when food is less abundant. An examination of the data in Table 7 reveals that specimens with a mean initial size of 323 μm fed daily reached reproductive maturity in 10 days with a standard deviation of ± 3 days. By contrast, specimens fed every 7 days reproduced on the average within 23 days, but the standard deviation is ± 15 days. Clearly, there is less synchrony in the latter group compared to the former group fed every day. The prolonged survival of slowly maturing individuals and their more distributed pattern of reproduction during periods of low food abundance is an adaptive feature that may enhance population stability and contribute to genetic continuity. Under oligotrophic conditions, a graded sequence of reproduction over time within a population of maturing organisms is more likely to permit survival of the offspring, given the limited food supply, than a massive nearly synchronized burst of reproduction. Moreover, if the unfavorable conditions worsen at any point in time resulting in massive death of offspring, there will still be sufficient mature individuals to contribute to continuity. Likewise, when prey are abundant, it is advantageous to mature rapidly, reproduce promptly and thus take advantage of the food abundance.

Among the species of radiolaria examined thus far, regulation of swarmer production appears to be more a function of endogenous physiological factors rather than environmental factors.[53] Within the limited research evidence available, it appears some radiolaria (e.g., *Spongodrymus* sp.) respond to favorable food supplies by living longer as observed also in the planktonic foraminifer *Hastigerina pelagica*. *H. pelagica* lacks closely associated algal symbionts, and among other significant factors, is clearly different from most radiolaria in morphology and physiology. The few radiolarian species we have examined, although rich in symbionts, do not appear to be dependent on symbiont-mediated light regulation of reproduction as occurs in *G. sacculifer*. On the whole, *G. sacculifer* appears to be considerably different in its physiological response to environmental signals for reproduction as compared to *H. pelagica* or many radiolaria. It is interesting to note that among the planktonic foraminifera, *H. pelagica* with a substantial bubble capsule exhibits a pronounced tendency toward zonal specialization as is also observed to a greater degree in radiolaria.[53]

Light-mediated control of reproduction in *G. sacculifer* is an intriguing phenomenon whose significance has not been fully elucidated. It is not immediately clear why prolonged darkness or low light intensity should result in early onset of swarmer production. One hypothetical explanation is that *G. sacculifer* is normally found in illuminated near-surface water where appropriate food is abundant and symbiont primary productivity is enhanced. If natural events should cause the foraminifer to be swept into deeper water, the early production of motile swarmers may be a mechanism to regain an advantageous position in the water column. Much additional research must be done on the comparative effects of daily photoperiod as well as light intensity and quality on the finer details of light regulatory mechanisms in reproduction of *G. sacculifer*. Further examination of other species is needed to determine if similar regulatory mechanisms exist and the possible mediating role of their symbionts in these processes.

Until we have more detailed information on the earliest stages of development in planktonic foraminifera and radiolaria, it will not be possible to trace the physiological development of the reproductive regulatory mechanisms nor to know at what point the symbionts first contribute substantially to host nutrition and influence reproductive maturation, as occurs in *G. sacculifer*. It is clearly essential to determine at what stage meiosis occurs, assuming some species have a sexual reproductive cycle, and to fully elucidate the complete life cycle of the planktonic foraminifera and radiolaria by identifying the missing link in development between swarmers and the second generation.

Moreover, much more research is needed on comparative processes of early ontogeny, maturation, and reproduction among marine planktonic protozoa to determine the relationship of reproductive strategies to their pattern of specialization. For example, to what extent is increased zonal specialization associated with greater efficiency in survival of swarmers and offspring, or in greater vigor or robustness during early ontogeny due to more specialized and differentiated cytoplasmic functions enhancing food apprehension, digestion, and coordinated metabolism.

The diversity of reproductive processes in planktonic foraminifera and radiolaria coupled with the complex interaction effects between host and symbiont provide a novel opportunity to investigate fundamental reproductive events within large, single-celled systems and thereby gain additional insights into the ecological, physiological, and evolutionary bases of specialization and reproduction in microzooplankton communities.

ACKNOWLEDGMENTS

I express sincere appreciation to my colleagues who over the years have participated in or encouraged aspects of this research, including Drs. A. W. H. Bé and N. R. Swanberg at Lamont-Doherty Geological Observatory of Columbia University, and Drs. Ch. Hemleben and M. Spindler of Tübingen University, West Germany, where I pursued some of the research on radiolaria during a sabbatical leave. S. Tuntivate-Choy, M. Botfield, J. Hacunda, D. Caron, and H. Spero were able research assistants during periods of the research summarized here. The staff at the Bermuda Biological Station (Ferry Reach, Bermuda), Bellairs Research Institute (St. James, Barbados), and the Spanish Research Institute (Teneriffe, Canary Islands) gave generous assistantship during phases of the field research. This research was supported in part by grants from the United States National Science Foundation, Biological Oceanography Division OCE 78-25450 and OCE 80-05131. This is Bermuda Biological Station Contribution No. 882 and Lamont-Doherty Geological Observatory Contribution No. 3618.

REFERENCES

1. Murray, J., On the distribution of the pelagic foraminifera at the surface and on the floor of the ocean, *Nat. Sci.*, 11, 17, 1897.
2. Rhumbler, L., Die foraminiferen (Thalamophoren) der Plankton-Expedition. Erster Teil: Die allegemeinen Organisationsverhaltnisse der Foraminiferen. *Plankton-Expedition der Humboldt-Stiftung, Ergebn.*, 3, 1, 1911.
3. Brandt, K., Die koloniebildenden Radiolarien (Sphaerozoeen) des Golfes von Neapel und der angrenzenden Meeresabschnitte. *Fauna und Flora des Golfes von Neapel*, 13, 1, 1885.
4. Brandt, K., Beitrage zur Kenntnis der Colliden. *Arch. Protistenkd.*, 1, 59, 1902.
5. Brandt, K., Zur Systematik der koloniebildenden Radiolarien. *Zool. Jahrb. Suppl.*, 8, 311, 1905.
6. Haeckel, E., Report on the Radiolaria collected by HMS "CHALLENGER" during the years 1873—76. Reports of the Scientific Research Voyage of the "CHALLENGER" 1873—1876. *Zoology*, 18, 3 parts, Neill and Co., Edinburgh, p. 1803, 1887.
7. Lee, J. J., Freudenthal, H. D., Kossoy, V., and Bé, A. W. H., Cytological observations on two planktonic foraminifera, *Globigerina bulloides* d'Orbigny, 1826 and *Globigerinoides ruber* (d'Orbigny, 1839) Cushman, 1927. *J. Protozool.*, 12, 531, 1965.

8. Cachon, J. and Cachon, M., Le système axopodial des Radiolaria nassellaires. *Arch. Protistenkd.,* 113, 80, 1971.

9. Cachon, J. and Cachon, M., Le systeme axopodial des Radiolaires Spaeroidés. I. Centroaxoplastidiés. *Arch. Protistenkd.,* 114, 51, 1972.

10. Cachon, J. and Cachon, M., Le système axopodial des Radiolaires Sphaeroidés. II. Les périaxoplastidiés III. Les cryptoaxoplastidiés (anoxoplastidies) IV. Les fusules et le système rhéoplasmique. *Arch. Protistenkd.,* 114, 291, 1972.

11. Cachon, J., Cachon, M., Febvre-Chevalier, C., and Febvre, J., Déterminisme de l'édification des systèmes microtubulaires stereoplasmiques d'Actinopodes. *Arch. Prostitenkd.,* 115, 137, 1973.

12. Hollande, A., Donnees ultrastructurales sur les isospores des radiolaires, *Protistologica,* 10, 567, 1974.

13. Hollande, A., Cachon, J., and Cachon, M., La signification de la membrane capsulaire des Radiolaires et ses rapports avec plasmalemme et les membranes du réticulum endoplasmique. Affinités entre Radiolaires, Heliozaires et Péridiniens. *Protistologica,* 6, 311, 1970.

14. Hollande, A. and Carre, D., Les xanthelles des radiolaires sphaerocollides, des acanthaires et de *Velellela velella:* Infrastructure-cytochimie-taxonomie, *Protistologica,* 10, 573, 1974.

15. Cachon, J. and Cachon, M., Constitution infrastructurale des microtubules du système axopodial des Radiolaires. Infrastructural Constitution of the Microtubuli of the Axopodial system of radiolaria, *Arch. Protistenkd.,* 120, 229, 1978.

16. Cachon, J. and Cachon, M., Nouvelle interprétation de la division nucléaire des Phaeodairés (Actinopodes). de M. Jean Cachon, M. Monique Cachon et M. Pierre Lecher, *C. R. Acad. Sci. Paris,* 276, 3311, 1973.

17. Cachon, J. and Cachon, M., Les processus sporogénétiques du Radiolaire, *Arch. Protistenkd.,* 111, 87, 1969.

18. Spindler, M. and Hemleben, C., Symbionts in planktonic Foraminifera (Protozoa), in *Endocytobiology, Endosymbiosis and Cell Biology,* Schwemmler, W. and Schenk, H. E. A., Eds., DeGruyter, Hawthorne, New York, 1980, 133.

19. Lipps, J. H. and Valentine, J. W., The role of foraminifera in the trophic structure of marine communities, *Lethaia,* 3, 279, 1970.

20. Casey, R., Gust, L., Leavesley, A., Williams, D., Reynolds, R., Duis, T., and Spaw, J. M., Ecological niches of radiolarians, planktonic foraminiferans and pteropods inferred from studies on living forms in the Gulf of Mexico and adjacent waters, *Trans. Gulf Coast Assoc. Geol. Soc.,* 29, 216, 1979.

21. Casey, R. E., Partridge, T. M., and Sloan, J. R., Radiolarian life spans, mortality rates, and seasonality gained from recent sediment and plankton samples, *Proceedings of the Second Conference: Roma, Edizioni Tecnoscienza Roma,* 1, 159, 1970.

22. Herring, P. J., Some features of the bioluminescence of the radiolarian *Thalassicolla* sp., *Mar. Biol.,* 53, 213, 1979.

23. Anderson, O. R., A cytoplasmic fine-structure study of two spumellarian radiolaria and their symbionts, *Mar. Micropaleontol.,* 1, 81, 1976.

24. Anderson, O. R. and Bé, A. W. H., The ultrastructure of a planktonic foraminifer *Globigerinoides sacculifer* (Brady) and its symbiotic dinoflagellates, *J. Foraminiferal Res.,* 6, 1, 1976.

25. Anderson, O. R., Cytoplasmic fine structure of nassellarian radiolaria, *Mar. Micropaleontol.,* 2, 251, 1977.

26. Anderson, O. R., Light and electron microscopic observations of feeding behavior, nutrition, and reproduction in laboratory cultures of *Thalassicolla nucleata, Tissue and Cell,* 10, 401, 1978.

27. Lee, J. J., McEnery, M., Pierce, S., Freudenthal, H. D., and Muller, W. A., Tracer experiments in feeding littoral foraminifera, *J. Protozool.,* 13, 659, 1966.

28. Anderson, O. R. and Bé, A. W. H., A cytochemical fine structure study of phagotrophy in a planktonic foraminifer, *Hastigerina pelagica* (d'Orbigny), *Biol. Bull.,* 151, 437, 1976.

29. Anderson, O. R., Spindler, M., Bé, A. W. H., and Hemleben, Ch., Trophic activity of planktonic foraminifera, *J. Mar. Biol. Assoc. U.K.,* 59, 791, 1979.

30. Bé, A. W. H., Hemleben, Ch., Anderson, O. R., Spindler, M., Hacunda, J., and Tuntivate-Choy, S., Laboratory and field observations of living planktonic foraminifera, *Micropaleontology,* 23, 155, 1977.

31. Anderson, O. R. and Bé, A. W. H., Recent advances in foraminiferan fine structure research, in *Foraminifera,* Vol. 3, Hedley, R. and Adams, C., Eds., Academic Press, London, New York, 1978, 121.

32. Anderson, O. R., Foraminifera, in *McGraw-Hill Yearbook of Science and Technology,* McGraw-Hill, New York, 1979, 205.

33. Anderson, O. R., Ultrastructure of a colonial radiolarian *Collozoum inerme* and a cytochemical determination of the role of its zooxanthellae, *Tissue and Cell,* 8, 195, 1976.

34. Anderson, O. R., Fine structure of a symbiont-bearing colonial radiolarian, *Collosphaera globularis*, and ¹⁴C isotopic evidence for assimilation of organic substances from its zooxanthellae, *J. Ultrastruct. Res.*, 62, 181, 1978.

35. Anderson, O. R., Radiolaria, in *Biochemistry and Physiology of Protozoa*, Vol. 3, 2nd ed., Levandowsky, M. and Hutner, S., Eds., Academic Press, New York, 1980, 1.

36. Anderson, O. R., Radiolarian fine structure and silica deposition, in *Silicon and Siliceous Structures in Biological Systems*, Simpson, T. and Volcani, B., Eds., Springer Verlag, Heidelberg, 1981, 347.

37. Anderson, O. R., Radiolarian symbiosis, in *Algal Symbiosis — a Continuum of Interaction Strategies*, Goff, L. and Lewin, J., Eds., Cambridge University Press, Cambridge, 1983, 69.

38. Anderson, O. R., Fine structure of a marine amoeba associated with a blue-green alga in the Sargasso Sea, *J. Protozool.*, 24, 370, 1977.

39. Bé, A. W. H., Hemleben, Ch., Anderson, O. R., and Spindler, M., Chamber formation in planktonic foraminifera, *Micropaleontology*, 25, 294, 1979.

40. Anderson, O. R. and Swanberg, N. R., Skeletal morphogenesis in some living collosphaerid radiolaria, *Mar. Micropaleontol.*, 6, 385, 1981.

41. Bé, A. W. H. and Anderson, O. R., Gametogenesis in planktonic foraminifera, *Science*, 192, 890, 1976.

42. Bé, A. W. H., Gametogenic calcification in a spinose planktonic foraminifer, *Globigerinoides sacculifer* (Brady), *Mar. Micropaleontol.*, 5, 283, 1980.

43. Ketten, D. R. and Edmond, J., Gametogenesis and calcification of planktonic foraminifera, *Nature*, 278, 546, 1979.

44. Swanberg, N. R. and Anderson, O. R., *Collozoum caudatum* sp. nov.: A giant colonial radiolarian from equatorial and Gulf Stream waters, *Deep-Sea Res.*, 28A, 1033, 1981.

45. Anderson, O. R., Fine structure of a collodarian radiolarian (*Sphaerozoum punctatum* Müller 1858) and cytoplasmic changes during reproduction, *Mar. Micropaleontol.*, 1, 287, 1976.

46. Hollande, A. and Martoja, R., Identification du cristalloide des isospores de Radiolaires a un cristal de celestite (Sr SO₄). Determination de la constitution du cristalloide per voie cytochimique et à laide de la microsonde électronique et du microanalyseur à émission ionique secondaire, *Protistologica*, 10, 603, 1974.

47. Bé, A. W. H., Caron, D. A., and Anderson, O. R., Effects of feeding frequency on life processes of the planktonic foraminifer *Globigerinoides sacculifer* in laboratory culture, *J. Mar. Biol. Assoc. U.K.*, 61, 257, 1981.

48. Spindler, M., Anderson, O. R., Hemleben, Ch., and Bé, A. W. H., Light and electron microscopic observations of gametogenesis in *Hastigerina pelagica* (Foraminifera), *J. Protozool.*, 25, 427, 1978.

49. Spindler, M., Hemleben, Ch., Bayer, U., Bé, A. W. H., and Anderson, O. R., Lunar periodicity of reproduction in the planktonic foraminifer *Hastigerina pelagica*, *Mar. Ecol. -Prog. Ser.*, 1, 61, 1979.

50. Caron, D. A., Bé, A. W. H., and Anderson, O. R., Effects of variations in light intensity on life processes of the planktonic foraminifer *Globigerinoides sacculifer* (Brady), *J. Mar. Biol. Assoc. U.K.*, 62, 435, 1982.

51. Bé, A. W. H., Spero, H., and Anderson, O. R., The effects of symbiont elimination and reinfection on the life processes of the planktonic foraminifer *Globigerinoides sacculifer*, *Mar. Biol.*, 70, 73, 1982.

52. Fairbanks, R. G. and Wiebe, P. H., Foraminifera and chlorophyll maximum: vertical distribution, seasonal succession, and paleooceanographic significance, *Science*, 209, 1524, 1980.

53. Anderson, O. R., *Radiolaria*, Springer-Verlag, New York, 1983, 355.

Chapter 4

PLANKTONIC COPEPODA (INCLUDING MONSTRILLOIDA)*

Charles C. Davis

TABLE OF CONTENTS

*   Contribution of the Marine Sciences Research Laboratory, Numbeı 481. Preparation was supported in part by Grant A-5074 from the National Science and Engineering Council of Canada.

I. INTRODUCTION

Copepods are small crustaceans found in marine and freshwater habitats. Most species in the orders Calanoida, Cyclopoida, and Harpacticoida are free-living planktonic or benthic forms, although parasitic forms occur in the order Cyclopoida. In the order Monstrilloida the larval forms are parasitic in invertebrate hosts and the adults are free-living. Several other orders are entirely parasitic. Free-living copepods may be herbivores, carnivores, detritivores, or omnivores. Herbivorous copepods can dominate the zooplankton in numbers and biomass and provide a major link between the primary producers, phytoplankton, and higher trophic levels. Many invertebrates, larval and adult fish, including basking and whale sharks, and whalebone whales, utilize copepods as a food source. Copepod fecal pellets also serve to transfer organic material to higher trophic levels in the benthic community. This transfer is especially important to abyssal communities which depend on "marine snow" for survival.[1]

II. GENERAL LIFE HISTORY

Most investigations indicate that, among the majority of planktonic Calanoida, Cyclopoida, and Harpacticoida, the life history includes common features. All species reproduce sexually although parthenogenesis occurs in the calanoid *Mormonilla*. Most species produce subitaneous eggs, which develop immediately; some species also produce resting eggs. Eggs hatch as nauplii with three pairs of cephalic appendages, the first and second antennae and the mandibles, on an unsegmented body (Figure 1). There are six naupliar stages, each separated by a molt. The stages are designated nauplius I to nauplius VI (N-I to N-VI). In *Pseudodiaptomus coronatus* there are only five naupliar stages.[3] Matthews[4] recorded only four naupliar stages for *Chiridius armatus*. Rudiments of additional cephalic appendages appear during the later naupliar stages (N-IV to N-VI) and these stages are properly designated as metanauplii. Stage N-VI molts into a copepodid in which all the cephalic appendages are well developed, two pairs of thoracic appendages appear, and the body shape resembles the adult form with the usual metasomal and urosomal divisions. As in the nauplius, there are five molts, which give rise to six distinct copepodid stages, namely copepodid I to copepodid VI (C-I to C-VI), C-VI being the adult stage. In many species the distinction between males and females can easily be detected as early as the copepodid IV, but in others no easy distinction can be made until the adult stage. Except in non-feeding naupliar stages, in those species where these occur, there is a progressive increase in size and substance from the first nauplius to the adult.

The life cycle of monstrilloid copepods is very different[5] (Figure 2). Only adults and nauplii occur in the plankton. Adults lack mouthparts and have only a rudimentary digestive tract. Eggs hatch into a nauplius whose basic structure is typical, except that there is no digestive system and grasping hooks occur on the second antennae and the mandibular palps. The nauplius seeks a host, which is usually a polychaetous annelid, but may be a gastropod or other marine invertebrate. It burrows into the host and upon entry, sheds its cuticle becoming a naked mass of cells which, in polychaetes, migrates to the dorsal vessel. A spinous sheath is produced around this larva, and anteriorly there is a pair of long antenna-like processes, through which nutrients are absorbed from the host. Within the sheath, development takes place up to the fully mature adult. The adult burrows out of the host by use of the spinous sheath, which is shed upon emergence (see Figure 2). The life history of monstrilloids has been so poorly studied that there are separate taxonomies for parasitic larval stages and planktonic adults. Adults almost always are rare in plankton samples presumably because they remain in the water column only long enough to reproduce.

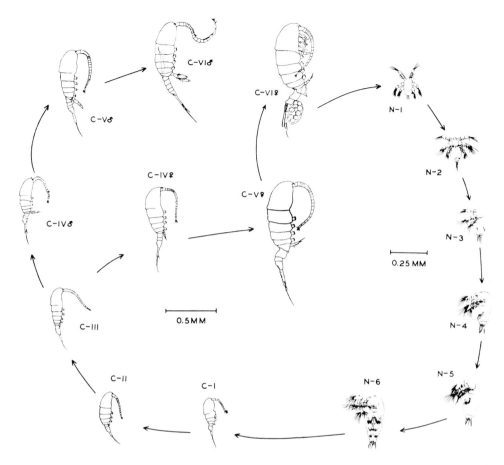

FIGURE 1. Generalized life history of a calanoid copepod, using *Eurytemora affinis* as an example. (Adapted from Katona.[2])

III. REPRODUCTION

The genetic mechanism for sex determination in planktonic copepods remains unclear, though Heberer[6] thought he detected heterogametous chromosomes in the male. The determination of sex ratios presents difficulties for several reasons: (1) males and females may develop at different rates; (2) adults of the sexes often have very different life expectancies, as in *Calanus finmarchicus*[7] and *Pseudocalanus* sp.;[8] (3) males and females may differ in their vertical or horizontal spatial distributions, e.g., predominantly male swarms at the surface in *Epilabidocera amphitrites*;[9] or (4) food, pollutants or other environmental factors may selectively alter the survival of one sex or perhaps cause sex reversals.[7,10] It is probable that the clutches of most copepod species start with approximately equal numbers of the two sexes. Equal sex ratios in copepodid IV and V stages are found in *Pseudocalanus minutus, Microcalanus pygameus, Acartia clausi*,[11] and *Labidocera diandra*.[12] Equal sex ratios for adults occur in non-swarming *Epilabidocera amphitrites*.[9] *Acartia clausi*.[11] and *A. tonsa*.[13] In *Centropages hamatus, Temora longicornis*,[12] and *Eurytemora affinis*,[13] adult males are more abundant than females. However, in most species the number of adult females exceeds males mainly because males have a shorter life span.[7,8,14,15]

There has been surprisingly little study of the mating process in planktonic copepods, although the external anatomy of many of the secondary sex structures involved

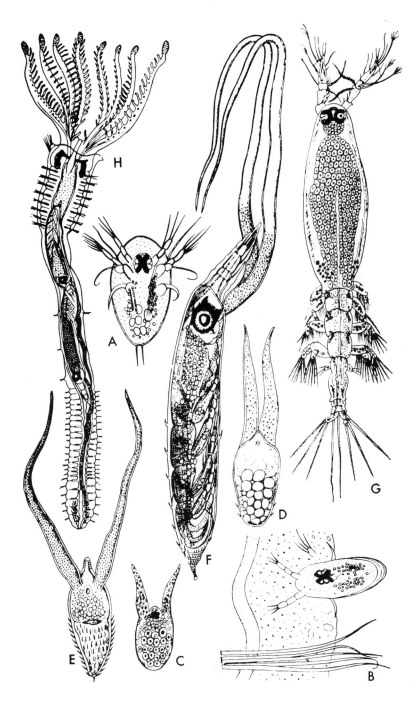

FIGURE 2.   Life history of a monstrilloid copepod, *Haemocera danae* (=*Monstrilla* sp.). (A) Nauplius; (B) nauplius entering its annelid host; (C to E) early developmental stages within the host; (F) copepodid ready for emergence from host; (G) planktonic adult female; (H) two well-developed monstrilloids within an annelid host. (From Baer, J. G., *Ecology of Animal Parasites,* University of Illinois Press, Champaign, 1951, 1. With permission.)

in the copulation process have been described in considerable detail because of their taxonomic importance. In most calanoids, one of the first antennal pair is modified in the male as a grasping organ and is termed geniculate. The modifications can range from moderate to extensive, while in a minority of forms, such as *Calanus*, there is no difference between the right and left antennae. In cyclopoids, monstrilloids, and harpacticoids, both the right and left first antennae are geniculate. In these three orders the male fifth thoracic appendages are similar or identical to those in the female. In most calanoid species both members of the fifth pair of thoracic appendages in the male are highly modified and are associated with grasping the female and transfer of the spermatophore. The modifications in various genera differ considerably and the function by no means can be the same in all species. In a few forms, such as *Calanus*, the modification is relatively slight and may involve only one member of the pair. In other genera both members of the pair are extremely different from the remainder of the thoracic appendages, and the left and right legs differ greatly from one another. In contrast, the first antennae of females of these four orders are not modified. In many species the female fifth legs are reduced in size and they may be lacking altogether. In harpacticoids, the female fifth legs are often plate-like and help shield the egg sac.

The behavior during copulation has been best described by Blades and Youngbluth[16,17] for *Labiodocera aestiva*, and by Blades[18] for *Centropages typicus*. Gauld[19] has given a less complete description for *Centropages hamatus, Temora longicornis, Eurytemora velox* and *Acartia clausi*. Katona[20] has studied some aspects of copulation in *Eurytemora affinis*. Attraction of the male to the female is probably through the chemoreception of pheromones[8,21,22] or by a combination of chemoreception and the detection of mechanical stimuli.[17,23] In the presence of females, calanoid males have been described as swimming in a looping and somersaulting fashion, very different from feeding behavior or escape responses. Presumably this gives them better opportunity to make contact with the females. The mating ritual basically involves grasping the female and transfer of the spermatophore. The method of grasping, the orientation of the male and female, the timing of spermatophore extrusion and the actual spermatophore transfer varies between species (Figure 3). This stylized mating behavior helps prevent hybridization.

Mating in marine cyclopoids, monstrilloids and harpacticoids has not been described in any detail to date. In one harpacticoid genus, *Harpacticus*, males grasp immature females (C-IV or V) and transfer the spermatophore during the molt of the female to the adult stage.[24-26] The close proximity of a male at the time of the female molt to the adult insures successful copulation. Freshwater cyclopoids have been studied by Wolf[27] and Hill and Coker.[28] A description of mating in freshwater harpacticoids has been given by Wolf.[27] According to Wolf, the thoracic appendages do not participate in the transfer of the spermatophore to the female, but the transfer occurs by direct juxtaposition of the two genital segments. Hill and Coker, however, observed use of all of the first four thoracic pairs in *Cyclops*. The appendages are used first to help dispose of the molt when the female undergoes ecdysis and then to carry the spermatophore to the female genital segment.

Copepod spermatophores vary somewhat from species to species, though the general structure and function appears to be similar in those studied to date[7,29-34] (Figure 4). The function of the spermatophore is to transfer the sperm to the female where it is stored in the seminal receptacle. In species where the male is short-lived, the contents of the single spermatophore is sufficient to fertilize the entire egg production of a single female. Normally only one spermatophore is deposited per female. However, when there is overcrowding or when mating competition is intense, more than one male may mate with a female and result in multiple spermatophores.[3,22,35,36] The spermatophores

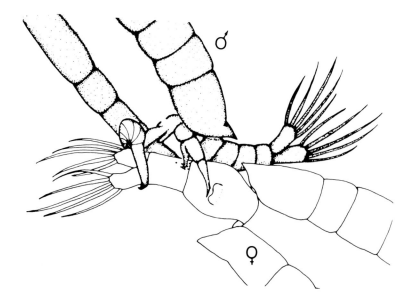

FIGURE 3.   Ventral view of copulation in *Labidocera aestiva.* The male has grasped the furcae of the female with the chela of his right fifth leg and is stroking a pore-field on the abdominal segment of the female, just posterior to the genital segment with his left leg. (From Blades, P. I. and Youngbluth, M. J., in *Evolution and Ecology of Zooplankton Communities,* Kerfoot, W. C., Ed., University Press of New England, Hanover, 1980, 39. With permission.)

may provide an additional barrier to hybridization in those species where the attachment end forms a coupling device which matches the genital field of the female.

Often a single female produces several consecutive batches of eggs. In some species, the fertility of consecutive broods declines and the female must be re-inseminated to renew production of fertile eggs.[13,35,37] Inasmuch as males, on the other hand, often have a short life span as adults and may appear only early in the breeding season, a single mating must serve to fertilize eggs over a more or less extended time. In the benthic harpacticoid, *Harpacticus obscurus,* a female can produce up to 14 fertile broods after one insemination.[26] Davis[38] found that *Acartia longiremis* females store viable sperm over winter (5.5 months) and produce fertile broods in spring.

Fecundity of individual females of a few species has been estimated. Most estimates have been determined in the laboratory; measurements of population fecundities in the field are less common. Fecundity varies with species, food availability and quality, presence of males, season, geographic locality, and temperature.[7-9,11,39-51] Laboratory estimates of fecundity may or may not reflect natural production. In general, larger-sized species produce more eggs than smaller species, and the period of egg production in high-latitude species is more seasonally restricted than in low-latitude species.

In some northerly regions, *Calanus finmarchicus* goes through only one generation a year, breeding in spring,[52,53] but in warmer waters there are two or more generations annually.[7,15] Other copepod species exhibit similar differences in the number of generations, depending upon latitude. In Tanquary Fjord, Ellesmere Island, *Pseudocalanus* breeds during the open-water season, but it takes two years for attainment of maturity.[54] In eastern Greenland waters the time for development to adulthood takes only one year,[55] while there appear to be six generations per year in Loch Striven, Scotland.[11,56] According to Cairns,[54] it takes "several years" in Tanquary Fjord for *Calanus glacialis* and *C. hyperboreus* to attain maturity.

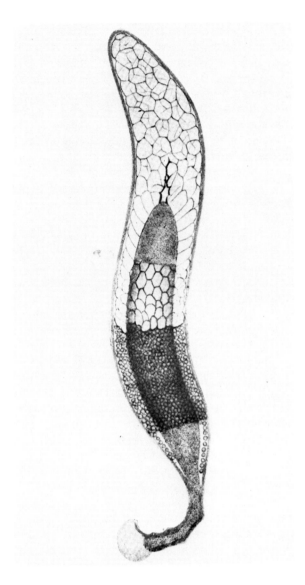

FIGURE 4. Spermatophore of *Candacia armata*, showing the distal swollen, distorted spermatozoa (Q-cells) and the proximal functional spermatozoa (B-cells), with a central core of secretion. (From Heberer, G., *Biol. Zentralbl.*, 74, 555, 1955. With permission.)

Other high-latitude species, such as the cyclopoid *Oithona similis*, appear to breed almost continuously, though more slowly in colder temperatures, dependent only on the amount of food available.[53] Thus, in far northern waters, where no primary production by photosynthetic phytoplankton occurs during the long winter night, breeding ceases entirely for a while,[38,55,56] whereas farther south, where at least some phytoplankton is present even in winter months, a number of the females carry egg sacs at all times.[11,15]

As summarized by Heinrich,[53] continuous breeding also is characteristic of tropical planktonic copepods. In temperate seas, the breeding cycles of planktonic copepods vary greatly. In cooler temperate regions breeding often is similar to that occurring in higher latitudes, whereas nearer to the equator continuous breeding is more frequent. During the warmer months of the year, when breeding is essentially continuous, it is possible to detect a number of cohorts. This was the basis of the interpretation by McLaren,[56] from data provided by Marshall[11] at Loch Striven, that there were six cohorts from March through August for *Pseudocalanus*. Johnson[57] also found six cohorts for *Acartia californiensis* between June and October in Yaquina Bay, Oregon.

Various authors have examined the rate of development of eggs or other stages in the development of copepods. As a rule this has been accomplished by culturing in the laboratory, in which case factors such as temperature and food supply can be closely controlled. Nevertheless, as usual in laboratory studies, there is considerable question concerning the validity of interpreting developments in nature on the basis of laboratory observations. Shallow-water temperatures in the field are seldom uniform, while food and other conditions may be sub-optimal. Furthermore, attempts to ascertain developmental rates in the open sea are complicated because one seldom can sample the same population from one observation period to another, and this leads to uncertainties. On the other hand, conditions in the laboratory, such as reduced turbulence in the cultures and increased bacterial contamination in small containers may affect growth rates considerably.

The development rate for eggs depends strongly upon the temperature, within the range the eggs can withstand (the range obviously being different for tropical species than for temperate and boreal species). Reported rates vary from 1 to 2 days for species of *Acartia* at high temperatures[58,59] and for *Eurytemora affinis*,[13] to several months when low (or high) temperatures enforce the eggs to enter a non-diapause dormancy.[59] As shown by McLaren,[60] there are small differences in the development rate at any given temperature of the eggs of *Pseudocalanus minutus* cultured at temperatures between 0°C and 15°C, even though they were taken from such different locations as Woods Hole in Massachusetts, Halifax in Nova Scotia, Milport in Scotland, Ogac Lake or Frobisher Bay, in the Canadian Northwest Territories. The same author, however, showed that there are distinct rate differences with temperature for differing copepod species including a large form of *Pseudocalanus* sp.

Even at the same temperatures, and in a single species, the rate of development may vary significantly from one season to another.[58,60] In a single species, the rate of development may be greater for larger eggs than for smaller ones, and the seasonal differences may well be based upon the production of eggs of differing sizes at different periods of the year, depending perhaps on differing ambient temperatures or on variations in the quantity and quality of food available for the females. On the other hand, the rate differences between species cannot be based solely on size. The large, copepod *Calanus glacialis* (4.5 to 5.5 mm), produces large eggs which develop distinctly faster than the much smaller eggs of *Pseudocalanus minutus*, including specimens obtained from the same arctic locality as *C. glacialis*.[60]

According to McLaren[60] and McLaren et al.,[61] the rate of development with temperature can be interpreted by use of the equation of Belĕhrádek:[62] $D = a(T - \alpha)^b$ in which D is the observed time in days for development, T is the temperature in °C, and a, α and b are "constants"; and a is the development time one degree above α, α is the temperature at which development time is infinitely long, and b reflects the slope of the curve. Landry[58] and Uye[59] have shown that at temperatures above a certain maximum, and before the attainment of lethality, the development rate, in some species at least, is slowed down. Other non-lethal, but non-optimal environmental conditions can retard the rate of egg development. High or low salinity,[59,63] darkness,[64] and low oxygen content of the water[59] can affect egg development rate. In general, however, these effects are much less extensive than the effects of varying temperature.

The rate of development of nauplii and copepodids depends upon the genetics of the species, the sex of the individual, and the temperature in which growth is occurring. The rate can differ between stages; earlier stages often develop more rapidly than later stages. Generation times (the time necessary for an egg to develop, hatch, and the resulting nauplius to mature to a reproductive adult) are longer for species in cooler and/or high latitude waters than for species in warmer and/or low latitude waters. Generation times can range from as little as 7 to 12 days for *Acartia tonsa* in temperate Chesapeake Bay[65,66] to as long as two years for *Pseudocalanus minutus* in the far northern waters off Ellesmere Island.[54] Within a species, development times can vary depending on geographic locality. McLaren,[56] using data from Marshall[11] for *P. minutus* populations from the more southerly Loch Striven, estimated development times ranging from 34.5 days at 8°C to 21 days at 21.5°C. Generation times in cold water species can be lengthened by delayed development of a stage or diapausal periods. In species such as *P. minutus* and *Calanus finmarchicus*, the long generation time is due to an overwintering generation of CIV or CV which enter a "resting" stage during the cold months.

In those stages where juveniles feed actively, the rate of development may be favorably or adversely affected by the quantity and quality of available food.[65] The development times for *Rhincalanus nasutus* fed on *Ditylum brightwelli* were 22 days at 15°C and 47 days at 10°C.[67] In contrast, the development times for this species when fed *Thalassiosira fluviatilis* were 35 days and 53 days respectively.

IV. EGGS AND HATCHING

Except in possible parthenogenic species, fertilization of eggs takes place during passage out of the female genital tract. Two types of eggs may be produced, namely the usual subitaneous eggs which usually develop immediately and diapausal eggs which require a dormancy period before hatching. Subitaneous eggs may fail to develop at once simply because they are produced at a time when conditions for development are unsuitable (e.g., low temperatures or low oxygen content of the water[68]), and these are capable of normal development at any time once suitable conditions are re-established. Grice and Marcus[69] proposed the term "quiescent" for this type of resting (dormant) egg. Diapausal eggs are produced in response to some environmental trigger(s) and cannot develop even under optimal conditions until the passage of a required dormancy period. To date, in the marine habitat, only calanoids are known to produce diapausal eggs. Both diapausal and quiescent eggs are found in marine sediments. Subitaneous eggs are laid singly into the water by some marine calanoids such as *Calanus*, *Acartia*, *Temora*, *Epilabiodocera*, and *Tortanus*. The females of other species carry egg sacs attached to the genital segment (e.g., *Eurytemora*, *Pseudocalanus*, *Euchaeta*, *Oithona*, and *Microsetella*). Cyclopoids bear two egg sacs (usually both with approximately the same number of eggs) while most calanoids and harpacticoids have only a single egg sac. Monstrilloid females hold their eggs between special paired (sometimes fused) spines on the ventral surface of the genital segment.

Dormancy in the eggs of marine copepods has received little attention until recently.[39,64,69-74] Grice and Marcus[69] state that dormancy has been shown for sixteen marine species but of these, experiments confirm true diapausal eggs for only six species (*Acartia californiensis*, *A. tonsa*, *Labidocera aestiva*, *Pontella meadi*, *P. mediterranea*, and *Tortanus forcipatus*). *Centropages hamatus* and *C. ponticus*[70] produce two types of morphologically distinct eggs. The type of eggs produced in the fall are present in the sediments and are probably diapausal eggs. *Tortanus forcipatus* and *P. mediterranea* also produce morphologically distinct diapausal eggs. Dormant eggs of the remaining species found in the sediments cannot be distinguished as diapausal or quiescent without knowledge of the conditions under which they were produced, the length of time in the sediments, and the environmental conditions prevailing during the dormancy period. The stimuli which trigger the production of diapausal eggs may include temperature,[39] abnormal salinity,[64] lowered food supply, crowding of adults, photoperiod,[73] or some combination of these. Temperature is an important environmental factor in determining the length of dormancy and the termination of diapause.

Both diapausal and quiescent eggs may be important in species succession in neritic and estuarine waters.[69,72] Individual physiological responses of different species to environmental variables could determine the type of egg produced, the length of dormancy, hatching, and the timing of these life history events over an annual cycle.

V. OVERWINTERING: DIAPAUSE AND DELAYED DEVELOPMENT

Diapause and other means of delayed development are adaptations for survival during deleterious conditions, particularly winter conditions in higher latitudes, but also other unfavorable situations such as high summer temperatures, low oxygen content of the water,[68] and lack of adequate quantity or quality of food. These mechanisms are best known for freshwater copepods, where they may affect eggs,[75] juvenile copepodids,[76-79] or adults.[80]

Dormancy in marine copepods has been studied more for the egg stage, than for more advanced life history stages. To my knowledge, no instances of true diapause in the naupliar stages of marine copepods have been reported, though Landry[81] has described an inhibition by darkness of hatching in *Acartia clausi* for otherwise fully developed eggs. Hence eggs of this species buried in bottom sediments would remain unhatched until again exposed to light upon becoming resuspended. Also, Coull and Dudley[82] have documented a delay in development of up to 30 days of a proportion of the hatched nauplii in certain marine (non-planktonic) harpacticoids, a condition that spreads out the timing of the attainment of adulthood.

In temperate-boreal areas, diapausal eggs serve as an overwintering stage and allow repopulation of areas after the return of favorable conditions in the spring. In species such as *Acartia tonsa* and *Labidocera aestiva*, adults disappear from the plankton and presumably die-off in the fall.[69] The population survives as diapausal eggs in the sediment and repopulate the plankton in the spring. The geographic distribution of these two copepods ranges from cold northern temperate latitudes to warmer southern latitudes. Populations from warm waters apparently do not produce diapausal eggs. This population variability suggests a genetic basis for diapause and may represent adaptation to local environments. This difference between populations may represent reduced gene flow and the beginning of reproductive isolation.[69]

Delayed development of juvenile copepodids has been reported frequently. Thus, overwintering of *Calanus finmarchicus* in C-IV and C-V has been described by Marshall and Orr[7] for the Firth of Clyde, and Davis[38] for the Tromsø region in northern Norway. That this is not a true diapause is suggested, however, by the observation[83] that these stages feed during the winter months, albeit at a greatly reduced level. A similar winter arrest of development in C-IV and C-V has been indicated for *Pseudocalanus minutus* in northern Norway.[38] The same was found by Grainger[84] at Igloolik in the Canadian Arctic (69° 20' N. Lat.), in east Greenland waters by Ussing[85] and Jespersen,[86] and in the open Atlantic Ocean on the Arctic Circle at Weather Ship "M" by Østvedt.[87] Retarded development at C-IV and C-V also has been intimated for *Pseudocalanus* in other far northern waters by Cairns[54] in Tanquary Fjord, Ellesmere Island, by McLaren[88] in Ogac Lake (a landlocked fjord), Baffin Island, and by Carter[89] in Tessiarsuk Lake, Northern Labrador (also a landlocked fjord).

Although none of the latter authors obtained winter samples, the species developed by the end of the growing season to advanced juvenile copepodites and was found in the adult stage when the water was ice-free again in May-June. This overwintering "strategy", however, applies to *Pseudocalanus* less and less as the populations occur farther to the south into warmer waters. Runnstrøm[90] in southwestern Norway reported females bearing egg sacs throughout the year (except December), while Marshall[11] described males, females, and females bearing spermatophores at Loch Striven, Scotland, even in January and February. Davis (unpublished) found ovigerous and/or spermatophore-bearing females throughout 14 months of investigation in Conception Bay, Newfoundland, although the numbers became very low in biological winter, especially in March. True diapause, however, evidently has not been documented for marine juvenile copepodid stages of copepods.

Unlike a number of species of freshwater cyclopoid and harpacticoid copepods, there are no reports of encystment of juveniles in the bottom sediments. Coull and Grant[91] have discovered encystment of adults of the non-planktonic marine harpacticoid, *Heteropsyllus nunni*, in bottom sands.

Certain species of planktonic adult copepods undergo a delayed development (undoubtedly non-diapausal) during the winter season. In the dark period of the winter above the Arctic Circle, Davis[38] has described the existence of only adult males and females for the planktonic harpacticoid, *Microsetella norvegica* (70° N. Lat., near Tromsø in Norway), while at the same locality, *Acartia longiremis* (Calanoida) occurred during the winter only as adult females, which began producing viable eggs in mid-March. Diapausal eggs have been demonstrated for some species of the genus *Acartia* by others (see above), and these also may have been produced by *A. longiremis* in the previous autumn, but this was not demonstrated. Farther south, both males and females of *A. longiremis* occur throughout the winter months. For example, in Conception Bay, Newfoundland, Davis[15] reported that, although the bulk of the rather meager winter population of this species were juvenile copepodids, males and females were present throughout the year, with adults increasing in numbers in March, while in Riga Bay in the Baltic Sea, Bodnek (cited in Heinrich[53]) stated that reproduction occurred in all months, including the winter.

VI. FOOD AND FEEDING

A. Naupliar Stage

As mentioned in the Introduction, the planktonic (adult and naupliar) stages of the monstrilloid copepods have no functional digestive system, and they depend upon nutrients stored during the long parasitic juvenile phase of life.

In a number of free-living planktonic copepods some of the early naupliar stages have no functional gut, and they therefore depend upon yolk or wax stored in the egg by the female for their source of nutrients. In *Rhincalanus nasutus*[92] and *Pontellopsis regalis*[93] the first naupliar stage does not feed. *Calanus finmarchicus*, *C. hyperboreus*, and *Euchaeta japonica* begin feeding in stage N-3.[94-97] Nicholls[98] and Bernard[99] indicated that *Euchaeta norvegica* and *E. marina* nauplii do not feed at any stage, and feeding does not begin until the first copepodite stage. Sekiguchi[100] pointed out that nauplii of *Aetidius*, *Candacia*, *Euchirella*, *Paraugaptilus*, and *Tortanus* lack a masticatory blade on the mandibles and may be incapable of feeding. Matthews[4] concluded that nauplii of *Chiridius armatus*, *Aetideus armatus*, and possibly *Xanthocalanus fallax* do not feed at any stage in western Norwegian waters. This may be due, in part, to stored wax reserves as in *Calanus plumchrus*.[101] The feeding strategies of copepod nauplii may relate to their latitudinal distribution, temperature regimes, and storage products.

Nauplii of planktonic copepods have been successfully cultured in the laboratory using diatoms, dinoflagellates, and other flagellates.[67,92,96,97,102-107] Herbivory appears predominant in calanoid nauplii, and to my knowledge there have been no definitive reports of carnivory among free-living copepod nauplii, even where the adults are predators. Lewis and Ramnarine[108] suggested that both the prefeeding and feeding naupliar stages of *Euchaeta japonica* are able to utilize dissolved organics as a nutrient source. This hypothesis requires further rigorous testing.

B. Copepodid Stages

The larger size of copepodid stages, especially of the adults, has led to more extensive investigations of feeding mechanisms and food. In the past, based primarily on work with *Calanus*[109] and freshwater *Diaptomus*,[102] planktonic copepods were considered to be primarily herbivorous and simple, indiscriminate filter feeders. Not surprisingly the picture presented in recent literature is considerably more complex.[110-150] Planktonic copepods may also be carnivores, omnivores, or detritivores. Selective feeding may occur even in herbivores through mechano- and chemoreceptors.

Filtration feeding has been reported to be widespread among calanoid copepods (*Calanus, Eucalanus, Temora, Centropages,* etc.), including among juvenile copepodids of most species. According to early studies, cephalic appendages set up feeding currents to carry suspended food particles to the mouth area where they were filtered out through a network of setae and subsequently ingested.[151-153] Recent work suggests that no calanoid in reality feeds in this manner. In fact, the structure of oral appendages in many calanoids would make such filtration virtually impossible. The feeding mechanism of cyclopoids and harpacticoids also is very different.

In calanoids, it has been shown that water currents are not carried into a filtration basket as previously thought, but instead bypass the maxillae and are carried laterally away from the copepod[110,111] (Figure 5). If a food particle is nearby, the second maxillae reach out very rapidly so that the particle is sucked into the space between them. The first maxillae also aid in synchronously pushing the particle into the feeding basket. Rapid closure of the second maxillae then forces the water out through a meshwork of setae and setules, trapping the food particle. Movements of the first maxillae endites result in scraping captured food particles off the second maxillae setae.[111-113] The mechanism by which the copepod detects the presence of the food particle prior to capture has not been clarified, but it is speculated that detection is chemosensory. The rejection or acceptance of a captured food particle is probably also based on chemoreceptors. Mechanoreceptors on the first antennae probably function in distant detection of moving prey but algal cell detection is probably entirely by chemoreceptors on the mouth parts.

It is assumed that food particles obtained as described above are phytoplankton cells, especially diatoms. Beklemishev[114,115] and Sullivan et al.[116] found the teeth of mandibular blades of herbivorous copepods are siliceous and well adapted to cracking the siliceous diatom walls. However, much of the available food in nature includes non-living detritus of various origins. Inorganic detritus obviously is unsuitable as food, though it can be and is, ingested. There has been discussion concerning the suitability (or lack thereof) of organic detritus as food, for obviously much of it is the most resistant and undigestible remains of previously living organisms or of their activities (lignins, cellulose, chitin, fecal pellets, etc.). Nevertheless, such detritus is subject to bacterial decay and if ingested, the attached bacteria have nutrient value. In discussing the food of deep-sea copepods, Harding[133] has concluded that those species which are not carnivorous must subsist mainly on the heterotrophic bacteria associated with detritus. Freely floating bacteria would be too small to be captured by the feeding

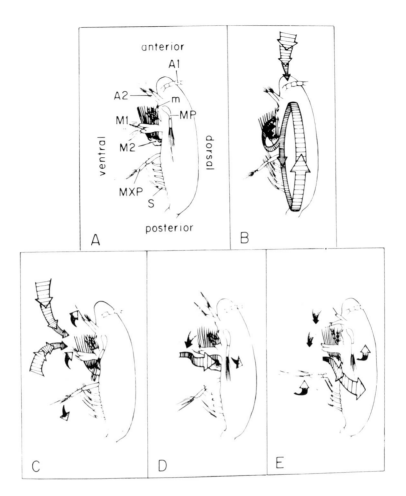

FIGURE 5.   Lateral view of feeding currents in *Eucalanus pileatus* as determined by Koehl and Strickler[111] (diagrammatic). Currents are shown by striped arrows, while movements of the appendages are indicated by stippled arrows. Movements away from the observer are depicted by a decreasing diameter of the shaft of the arrow, and movements towards the observer by an increasing diameter of the shaft. (A) Diagram of the stylized position of the appendages: A_1, A_2 = first and second antennae; m = mouth; M_1, M_2 = first and second maxillae; MP = mandibular palp; MXP = maxilliped; S = swimming legs. (B) A_2 and M_1 beat forward and dorsally, while MXP beats posteriorly and dorsally to produce a current that flows towards m and M_2. (C) The current is accentuated and diverted towards the observer when M_1 beats posteriorly and ventrally and MP beats towards the dorsal surface. (D) The feeding current bypasses the oral structures when A_2 and M_1 beat posteriorly and dorsally, MXP beats forward and ventrally, and MP beats anteriorly and ventrally. (From Koehl, M. A. R. and Strickler, J. R., *Limnol. Oceanogr.*, 26, 1062, 1981. With permission.)

devices of copepods. Arashkevich[134,135] has concluded that most copepods living deeper than 4000 m are predominantly mixed filter- and predatory feeders.

Most cyclopoids have uniramous second antennae and there is no trace of any true filtration mechanism in the oral region. They therefore must obtain their food by grasping individual particles or by attacking (or "parasitizing") larger prey.[122] Some species, such as *Oithona similis*, are mainly herbivorous, but the structure of the cyclopoid mouth parts has allowed easy evolution to parasitism and carnivory. The second maxillae and maxillipeds, instead of being filtering structures and devices for directing

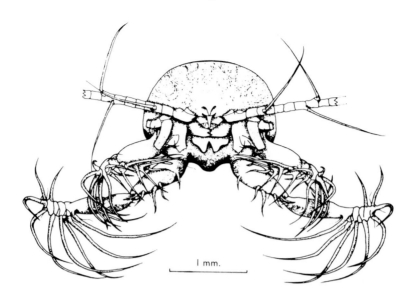

FIGURE 6.   Portrait of a *Paraeuchaeta* sp. female from an anterior view to emphasize the nature of the mouthparts in a predatory copepod. (From Wickstead, J. H., *Proc. Zool. Soc. Lond.*, 139, 545, 1962. With permission.)

food particles into a filtration chamber, are supplied with heavy setae suitable for grasping animal prey. Also in the Siphonostoma (*Oncaea* and *Corycaeus* among planktonic genera), the lips tend to be drawn out into a siphon suitable for sucking the liquid contents of prey. Little appears to be known concerning the feeding mechanisms in planktonic harpacticoids.

Carnivory occurs in cyclopoids and calanoids. Some larger calanoids may be entirely carnivorous and their mouth parts are modified for seizing large, active prey such as fish larvae, chaetognaths, or other copepods (Figure 6). We might expect carnivorous copepods to feed in a fashion similar to other "size-dependent predators" as described for freshwater carnivores by Zaret.[140] Predation by "size-dependent" carnivores on smaller prey is inefficient because of the time required to obtain and ingest enough prey. Efficiency increases with increasing size of prey until the prey becomes so large that handling time to subdue and ingest it leads to too much expenditure of energy. At this point, efficiency decreases again. Graphing efficiency against prey size produces a bell-shaped electivity curve. Studies on calanoid and cyclopoid carnivores have provided examples from all sections of the curve. Båmstedt and Holt[141] provided *Euchaeta norvegica* with three different prey sizes. *Euchaeta norvegica* preferentially selected the smallest prey species and consumption rates for the two larger species were towards the right side of the bell shaped curve. *Labidocera trispinosa* also feeds mainly on small prey, such as copepod nauplii.[139] Observations of predation on large prey have been mainly qualitative and have included young anchovy larvae[140] and chaetognaths.[143,144] Wickstead[122] described and figured two cyclopoid genera (*Corycaeus* spp. and *Oncaea* spp.) holding and presumably feeding on large chaetognaths, appendicularians and the copepod, *Rhincalanus*.

Prey detection is presumed to be by chemo- and mechanoreception as in freshwater copepods.[148-150] A few species (*Corycaeus*, *Monstrilla*, and *Anomalocera*) have well developed eyes with lenses, but the eye structure does not appear to allow the detection of images (although to my knowledge this has never been tested). Most species have a small poorly developed eye spot(s). We have insufficient knowledge concerning the impact of copepod carnivory in the marine ecosystem. Emphasis in quantitative studies has been almost entirely on herbivorous feeding, partly because of the outmoded concept that all copepods are primarily herbivorous, and partly because quantitative estimates of carnivory are much more difficult to undertake.

The classification of a copepod as a herbivore, carnivore, detritivore, or omnivore presents problems. Various authors have attempted to categorize copepods based on mouth part structure and supposed function.[118,119] However there is overlap in feeding patterns. Copepods, such as *Pontella meadi* which are primarily carnivorous, will seize and feed on diatoms. Crustacean larvae have been observed in the gut contents of herbivorous *Calanus*. *Acartia* feeds on algal cells as well as large animal prey, indicating more than one type of feeding mechanism.[117,144] Turner[154] recently has shown from the fecal pellets of two calanoid species and sampling of the concurrent plankton community, that these copepods were opportunistic feeders and fed on a variety of phytoplankters, tintinnids, crustaceans, and detritus.

In non-feeding stages, storage products of fat or wax are the source of metabolic energy.[101] Fat is distributed throughout the body and is used as a short-term metabolic fuel. Wax is stored in sacs and is used to meet long term energy requirements. Both fat and wax have similar densities, caloric value, and compressibility. Wax, however, becomes more buoyant when warmed. This characteristic may be important to vertically migrating copepods. Temperate, boreal, and deepwater copepods store large accumulations of wax to provide an energy source during overwintering or sparse food periods. Copepods in warmer waters accumulate very little wax. Non-feeding nauplii, copepodids, and adults which rely on wax reserves often have reduced mouth parts.[7,98,116,121]

VII. VERTICAL DISTRIBUTION AND MIGRATION PATTERNS

Various species of planktonic copepods occupy certain depths in the water column. These depths, or changes in depths, can be influenced by age, sex, reproductive state, endogenous rhythms, feeding strategies, light intensity and spectrum, presence of predators, temperature, day length, and other biological and physical factors. The regulation of spatial and temporal distribution can confer adaptive advantages through increased isolation or conversely through increased encounters with food, predators, competitors, and mates. Gerritsen[23] stated "Rates and probabilities of these encounters influence biological interactions and can therefore play a large part in the evolution and distribution of organisms." Current hypotheses on depth distributions and migrations, such as the resource utilization hypothesis and the predator avoidance hypothesis, directly involve rates and probabilities of encounters. Such encounters could even substantially alter the gene frequency within planktonic copepod populations through selective pressures of visual predators or through isolation of populations and inbreeding.

Some calanoids remain in the uppermost layers of water all their lives, e.g., pontellids;[155-157] others occur at varying depths with no life cycle stages occurring near the surface, e.g., *Microcalanus pygmaeus*[11] and *Paraeuchaeta rubra*;[100] while others migrate vertically. Vertical migrations or displacement can be of three kinds, diel vertical migrations, ontogenetic movement, and seasonal movements. Ontogenetic movements often are seasonal and, according to Longhurst,[158] a separation is not warranted.

Ontogenetic vertical movements occur in a number of species, for example in the genera *Calanus*, *Eucalanus*, *Calanoides*, and *Rhincalanus*. *Calanus cristatus*[100] lays its eggs below 500 m, where the nauplii hatch. The nauplii then swim upward to the epipelagic region, so that after metamorphosis, stages C-I to C-IV occur mainly above

200 m. In contrast, C-V inhabits all depths down to 2000 m, but this varies with the season. In the spring C-V is found mainly above 50 m, in the summer there is a bimodal distribution and C-V lives between 200 and 300 m or between 1500 and 1700 m; in winter the vertical distribution is again bimodal, mainly above 100 m and between 800 and 1000 m. The adults occur mainly below 500 m, where spawning occurs. *Calanus finmarchicus* likewise exhibits ontogenetic vertical movements.[7] Eggs and nauplii occur mainly in the upper layers of the water, C-I and C-III live somewhat deeper, while C-IV and C-V migrate in winter to much deeper layers of the water (>100 m). Adults then migrate back to surface waters to spawn.

On the other hand, in *Eucalanus bungii bungii*,[100] the eggs are produced above 50 m, where the emerged nauplii remain until metamorphosis. C-I to C-IV occur mainly below 200 m, but as in *Calanus* stage C-V has a much wider depth range, for it occurs at all depths to 2000 m. Again as in *Calanus*, the depth distribution of C-V varies with the season. In spring most specimens are to be found above 100 m, whereas in summer they occur between 100 and 300 m. In winter the bulk of the population inhabits deeper water between 400 and 600 m.

These ontogenetic vertical movements have to be viewed as *population* movements[159] (Figure 7), rather than as individual movements, because a number of individuals do not fit the described pattern of vertical distribution, especially in shallower waters.[15,36] Occasionally, large swarms of juvenile (C-IV and C-V) *Calanus finmarchicus* occur at the very surface at midday, for unexplained reasons[160] (also Davis in Kaldfjorden, northern Norway, unpublished).

Sekiguchi[100] contends that ontogenetic vertical movements of the type discussed above are characteristic of cold-water species that are primarily herbivorous. There would appear to be adaptive advantages in high-latitude locations for the nauplii to live in the upper illuminated layers of the water where their food is more abundant.

Diel vertical migrations have been studied for many years. Although there is no question concerning the reality of these daily movements in many species, the propensity of copepods to aggregate into smaller or larger swarms at various depths, movements of water masses, and the inefficiency of collection methods always introduce a degree of uncertainty into the results. A number of investigations have been undertaken in the laboratory, but as a rule the laboratory conditions (the nature of artificial light sources, reflections from the walls of containers, unnatural food conditions, and lack of turbulence) are so dissimilar to those found in nature that results applicable to naturally occurring populations, although interesting, are in some doubt.

Older reviews on diel migrations can be found in Russell,[161] Kikuchi,[162] Cushing,[163] Bainbridge,[164] Banse,[165] and Vinogradov.[166] Physico-chemical environmental factors (temperature, salinity, light quality and quantity, etc.) and biological environmental factors (available food, etc.), as well as the age, physiological condition, and reproductive stage of the copepods, may disrupt or even reverse the pattern of diel vertical migration; however, the usual pattern is as illustrated in Figure 8. At night, the population moves upward towards, or even to, the surface, while in the daytime the bulk of individuals moves downward into deeper water. Migrations can be meager or very extensive depending upon the species. In addition herbivores may be more responsive to migratory stimuli than omnivores, carnivores, or detritivores.[167]

Different ontological stages of copepods often migrate differently. Thus, the weaker swimming abilities of naupliar stages prevent them from making extensive migrations as do copepodites, especially adults. In *Calanus finmarchicus* at least at certain times of the year, C-V's do not migrate diurnally. Males migrate only somewhat, and female C-VI's do so extensively.[168] Ripe female *C. finmarchicus* migrated upward earlier and more extensively than those whose ovaries were still immature. Similarly, Hayward[169]

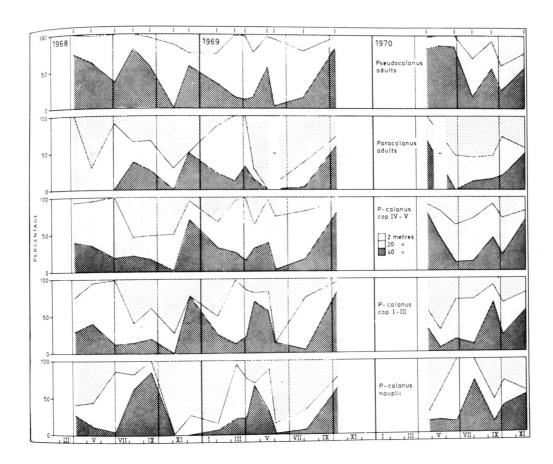

FIGURE 7. Ontogenetic vertical movements of populations of *Paracalanus parvus* in percent of populations occurring at 2, 20 and 40 m . (From Eriksson, S., *Zoon*, 1, 37, 1973. With permission.)

showed that males of *Pleuromamma piseki* with well-developed spermatophores migrated closer to the surface than those that were less mature. Thus, they would come closer to the vicinity of ripe females. However, males with spermatophores of the species *Scolecithrix danae* and *Undinula darwini*, which are weak migrators, did not exhibit such depth differentials.[169] When males and females migrate to different levels during the daytime or nighttime, as occurs in *C. finmarchicus* and in many other species, the two groups are separated into water where currents differ in direction, and/or in strength. In this way inbreeding within populations is minimized, and genetic mixing is maximized. This would contribute to the cosmopolitanism (or at least the widespread occurrence) that is evident in many species. Bainbridge[164] listed 49 species of marine planktonic calanoids and four species of cyclopoids that have been reported to make diel vertical migrations.

There has been considerable speculation and some experimental work designed to elucidate questions concerning both (1) the benefits to those species that migrate and (2) the stimuli that initiate the migrations. To date, controversy still surrounds the answers. Perhaps much of the disagreement comes from differences in stimuli and behavior among the many species involved — a matter that most authors have tended to ignore (e.g., in Davis[170]).

The most obvious possible adaptive value of downward migration during daylight hours is that copepods tend to avoid predators that feed visually.[8,171] During darkness,

FIGURE 8.   The "classical" picture of diel vertical migration, from observations on *Calanus finmarchicus.* (From Nicholls, A. G., *J. Mar. Biol. Assoc. U.K.,* Cambridge University Press, 19, 139, 1933. With permission.)

the copepods then rise upward to levels that are richer in food, and in which darkness reduces visual predation of plankton. The reduced size and lesser conspicuousness of nauplii allows them to remain in the richer food levels without, or without extensive, migration. Furthermore, the strength of sunlight in the upper levels can be detrimental to the animals if they remain near the surface in daytime.

McLaren[172] suggested that daytime migration to deeper, colder waters might be adaptive by reducing metabolic rates and conserving energy. Such metabolic reduction would be even more important in long-term survival of overwintering stages of temperate and boreal copepods. Enright[173] went further and suggested that there was nutritional advantage in waiting without feeding through daylight hours because during the daytime, photosynthesis would enhance the amount of organic material incorporated into the phytoplankton. To test this hypothesis, Enright and Honegger[174] undertook field observations on *Calanus pacificus,* in three sampling programs each of three days duration. On those sampling days when phytoplankton was scarce, vertical migration occurred only after sunset, whereas on days when phytoplankton was abundant, the migration commenced before darkness. The authors speculated that when food is scarce the copepods do not migrate upward because the danger from predation was "of overriding importance", whereas when food was abundant avoidance of the predators "was relatively unimportant". It is not clear how copepods are able to determine the concentration of food above them in order to make their "decision". However, one possibility is that the absorption and scattering of phytoplankton change the light spectrum and that herbivores are keyed to such shifts and changes in intensity (e.g., see Herman,[175] Dingle[176]).

As to the stimulus(i) that initiates diel vertical migrations, whether up or down, the perferendum hypothesis that the animals follow a certain light intensity[177] has been challenged as has the hypothesis of negative phototaxis balanced by a weaker geotropism. Swarming of *Calanus finmarchicus* at the very surface on sunny days,[168] commencement of upward migration before sunset,[174] downward migration beginning as early as 1:00 a.m.,[160] lack of migration in Arctic zooplankton during periods of continuous daylight,[178] and lack of migration by all individuals in a population as shown in numerous publications, mitigate against any such simplistic explanations, pertaining to all species. Individual species may however, have their own characteristic responses. It is clear that the influence of light intensity and quality is important; the movements by-and-large are correlated with diurnal day-night events, despite their irregularities and ambiguities.

Circadian rhythms, with light as the Zeitgeber, have been suggested. The lack of migrations for otherwise migratory species in conditions of 24-hour daylight in far northern summers,[178,179] and the presence of vertical population shifts for the same species on dull days,[162,180] and on brightly moon-lit nights, suggests that this explanation also is an oversimplification. With freshwater *Cyclops* spp. Strickler[181] found under experimental conditions that such an endogenous rhythm "was only barely perceptible", and that the animals responded mainly to changes in light intensity.

Attraction upwards by hungry copepods, through the existence above of phytoplankton as a food source, coupled with negative phototropism, has been suggested a number of times. Bainbridge[182] found experimentally that *Calanus* swarm upward in the presence of several phytoplankton species. The nature of a possible stimulus, if any, coming to the copepods from the phytoplankton directly (e.g., ectocrines) or indirectly (e.g., shift in light spectrum) remains uncertain, for in nature at least, the vertical distances are as a rule too great. In recent experimental work, Bohrer[167] showed that in a 100 m³ aquatron tank (depth 10 m), the extent of the nighttime vertical movements of copepods was determined by the depth of enriched subsurface layers of phytoplankton. The copepods migrated to the dense layer, then remained in that position. Without the diatoms, the migration was still initiated at the same time and continued until the surface was reached by large numbers of the copepods.

REFERENCES

1. Turner, J. T., Microbial attachment to copepod fecal pellets and its possible ecological significance, *Trans. Am. Microsc. Soc.*, 98, 131, 1979.
2. Katona, S. K., The developmental stages of *Eurytemora affinis* (Poppe, 1880) (Copepoda, Calanoida) raised in laboratory cultures, including a comparison of the larvae of *Eurytemora americana* Williams, 1906, and *Eurytemora herdmani* Thompson and Scott, 1897, *Crustaceana*, 21, 5, 1971.
3. Jacobs, J., Laboratory cultivation of the marine copepod *Pseudodiaptomus coronatus* Williams, *Limnol. Oceanogr.*, 6, 443, 1961.
4. Matthews, J. B. L., On the biology of some bottom-living copepods (Aetideidae and Phaennidae) from western Norway, *Sarsia*, 16, 1, 1964.
5. Noble, E. R. and Noble, G. A., *Parasitology: the Biology of Animal Parasites*, Lea and Febiger, Philadelphia, 1961, 1.
6. Heberer, G., Die Spermatogenese der Copepoden. II. Das Conjugations-und Reduktionsprobleme in der Spermatogenese der calanoiden Copepoden mit einem Anhang über die Spermatogenese von *Sapphirina ovatolanceolata* (Dana), *Z. Wiss. Zool.*, 123, 555, 1932.
7. Marshall, S. M. and Orr, A. P., *The Biology of a Marine Copepod, Calanus finmarchicus (Gunnerus)*, Oliver and Boyd, Edinburgh, 1955, 1.
8. Corkett, C. J. and McLaren, I. A., The biology of *Pseudocalanus*, *Adv. Mar. Biol.*, 15, 1, 1978.
9. Park, T. S., The biology of a calanoid copepod, *Epilabidocera amphitrites* McMurrich, *La Cellule*, 66, 129, 1966.
10. Eriksson, S., The biology of marine planktonic Copepoda on the west coast of Sweden, *Zoon*, 1, 37, 1973.
11. Marshall, S. M., On the biology of the small copepods in Loch Striven, *J. Mar. Biol. Assoc. U. K.*, 28, 45, 1949.
12. Fleminger, A., Taxonomy, distribution, and polymorphism in the *Labidocera jollae* group with remarks on evolution within the group (Copepoda: Calanoida), *Proc. U.S. Natl. Mus.*, 120, 1, 1967.
13. Heinle, D. R., Population dynamics of exploited cultures of calanoid copepods, *Helgol. Wiss. Meeresunters.*, 20, 360, 1970.
14. Conover, R. J., Notes on the molting cycle, development of sexual characters and sex ratio in *Calanus hyperboreus*, *Crustaceana*, 8, 308, 1965.
15. Davis, C. C., A preliminary quantitative study of the zooplankton from Conception Bay, insular Newfoundland, *Int. Rev. Gesamten Hydrobiol.*, 67, 713, 1982.
16. Blades, P. I. and Youngbluth, M. J., Mating behavior of *Labidocera aestiva* (Copepoda: Calanoida), *Mar. Biol.*, 51, 339, 1979.

17. Blades, P. I. and Youngbluth, M. J., Morphological, physiological, and behavioral aspects of mating in calanoid copepods, in Kerfoot, W. C., Ed., *Evolution and Ecology of Zooplankton Communities,* University Press of New England, Hanover, 1980, 39.

18. Blades, P. I., Mating behavior of *Centropages typicus* (Copepoda: Calanoida), *Mar. Biol.,* 40, 57, 1977.

19. Gauld, D. T., Copulation in calanoid copepods, *Nature,* 180, 510, 1957.

20. Katona, S. K., Copulation in the copepod *Eurytemora affinis* (Poppe, 1880), *Crustaceana,* 28, 89, 1975.

21. Katona, S. K., Evidence for sex pheromones in planktonic copepods, *Limnol. Oceanogr.,* 18, 574, 1973.

22. Hopkins, C. C. E. and Machin, D., Patterns of spermatophore distribution and placement in *Euchaeta norvegica* (Copepoda: Calanoida), *J. Mar. Biol. Assoc. U.K.,* 57, 113, 1977.

23. Gerritsen, J., Adaptive responses to encounter problems, in Kerfoot, W. C., Ed., *Evolution and Ecology of Zooplankton Communities,* University Press of New England, Hanover, 52, 1980.

24. Williams, L. W., The significance of the grasping antennae of harpacticoid copepods, *Science,* 25, 225, 1907.

25. Jewett, S. C. and Feder, H. M., Biology of the harpacticoid copepod, *Harpacticus uniremis* Kroyer on Dayville Flates, Port Valdez, Alaska, *Ophelia,* 16, 111, 1977.

26. Walker, L. M., Reproductive biology and development of a marine harpacticoid reared in the laboratory, *J. Crust. Biol.,* 1, 376, 1981.

27. Wolf, E., Die Fortpflangungsverhältnisse unserer einheimischen Copepoden, *Zool. Jahrb. Abt. Syst.,* 22, 101, 1905.

28. Hill, L. L. and Coker, R. E., Observations on mating habits of *Cyclops, J. Elisha Mitchell Sci. Soc.,* 45, 206, 1929.

29. Heberer, G., Untersuchungen über Bau and Funktion der Genitalorgane der Copepoden. I. Der männliche Genitalapparat der Calanoiden Copepoden, *Z. Mikrosk.-Anat. Forsch.,* 31, 250, 1932.

30. Heberer, G., Weitere Ergebnisse über Bildung und Bau der Genitalorgane der Spermatophore und Spermatophorenkoppel-apparate bei calanoiden Copepoden, *Verh. Dtsch. Zool. Ges.,* 39, 86, 1937.

31. Heberer, G., Physiologischer Spermiendualismus bei harpacticoiden Copepoden, *Naturwissenschaften,* 42, 261, 1955.

32. Heberer, G., Die Spermatophoren der Corycaeidae (Cop.), *Biol. Zentralbl.,* 74, 555, 1955.

33. Hopkins, C. C. E., The male genital system, and spermatophore production and function in *Euchaeta norvegica* (Copepoda: Calanoida), *J. Exp. Mar. Biol. Ecol.,* 35, 197, 1978.

34. Lee, C. M., Structure and function of the spermatophore and its coupling device in the Centropagidae (Copepoda: Calanoida), *Bull. Mar. Ecol.,* 8, 1, 1972.

35. Katona, S. K., Copulation in the copepod *Eurytemora affinis* (Poppe 1880), *Crustaceana,* 28, 89, 1975.

36. Giesbrecht, W., Die freilebenden Copepoden der Kieler Foehrde, *Ber. Komm. Wiss. Unters. Dtsch. Meere, Kiel,* 4, 85, 1882.

37. Wilson, D. F. and Parrish, K. K., Remating in a planktonic marine calanoid copepod, *Mar. Biol.,* 9, 202, 1971.

38. Davis, C. C., Overwintering strategies of common planktonic copepods in some North Norway fjords and sounds, *Astarte,* 9, 37, 1976.

39. Johnson, J. K., Effects of temperature and salinity on production and hatching of dormant eggs of *Acartia californiensis* (Copepoda) in an Oregon estuary, *Fish. Bull.,* 77, 567, 1980.

40. Marshall, S. M. and Orr, A. P., On the biology of *Calanus finmarchicus.* VII. Factors affecting egg production, *J. Mar. Biol. Assoc. U.K.,* 30, 527, 1952.

41. Conover, R. J., Reproductive cycle, early development, and fecundity in laboratory populations of the copepod *Calanus hyperboreus, Crustaceana,* 13, 61, 1967.

42. Corkett, C. J. and McLaren, I. A., Egg production and oil storage by the copepod *Pseudocalanus* in the laboratory, *J. Exp. Mar. Biol. Ecol.,* 3, 90, 1969.

43. Harris, R. P. and Paffenhöfer, G.-A., Feeding, growth and reproduction of the marine planktonic copepod *Temora longicornis* Müller, *J. Mar. Biol. Assoc. U.K.,* 56, 675, 1976.

44. Bernard, M., Quelques aspects de la biologie du Copépode pélagique *Temora stylifera* en Méditerranée: essai d'ecologie expérimentale, *Pelagos,* 11, 1, 1970.

45. Razouls, S., Maturité sexuelle et fécondite chez les femelles de *Temora stylifera,* copépode pélagique (Copepoda Calanoidea), *Arch. Zool. Exp. Gen.,* 115, 387, 1974.

46. Paffenhöfer, G.-A. and Knowles, S. C., Laboratory experiments on feeding, growth, and fecundity of, and effects of cadmium on *Pseudodiaptomus, Bull. Mar. Sci.,* 28, 574, 1978.

47. Cahoon, L. B., Reproductive response of *Acartia tonsa* to variations in food ration and quality, *Deep-Sea Res.,* 28A, 1215, 1981.

48. Parrish, K. K. and Wilson, D. F., Fecundity studies on *Acartia tonsa* (Copepoda: Calanoida) in standardized culture, *Mar. Biol.,* 45, 65, 1978.

49. Landry, M. R., Population dynamics and production of a planktonic marine copepod, *Acartia clausii,* in a small temperate lagoon on San Juan Island, Washington, *Int. Rev. Gesamten Hydrobiol.,* 63, 77, 1978.

50. Bernard, M., Le cycle vital en laboratoire d'un Copépode pélaguique de Méditerraneé, *Euterpina acutifrons* Claus, *Pelagos,* 1, 35, 1963.

51. Valentin, J., La ponte et les oeufs chez les Copepodes du Golfe de Marseille: cycle annuel et etude experimentale, *Tethys,* 4, 349, 1972.

52. Kielhorn, W. V., The biology of the surface zone zooplankton of a boreo-arctic Atlantic Ocean area, *J. Fish. Res. Board Can.,* 9, 223, 1952.

53. Heinrich, A. K., The life histories of plankton animals and seasonal cycles of plankton communities in the oceans, *J. Cons. Perm. Int. Explor. Mer.,* 27, 15, 1962.

54. Cairns, A. A., The zooplankton of Tanquary Fjord, Ellesmere Island, with special reference to calanoid copepods, *J. Fish. Res. Board Can.,* 24, 555, 1967.

55. Digby, P. S. B., The biology of the marine planktonic copepods of Scoresby Sound, East Greenland, *J. Anim. Ecol.,* 23, 298, 1954.

56. McLaren, I. A., Generation lengths of some temperate marine copepods: estimation, prediction, and implications, *J. Fish. Res. Board Can.,* 35, 1330, 1978.

57. Johnson, J. K., Population dynamics and cohort persistence of *Acartia californiensis* (Copepoda: Calanoida) in Yaquina Bay, Oregon, *Ph.D. dissertation,* Oregon State University, Corvallis, 1, 1980.

58. Landry, M. R., Seasonal temperature effects and predicting development rates of marine copepod eggs, *Limnol. Oceanogr.,* 20, 434, 1975.

59. Uye, S., Development of neritic copepods *Acartia clausi* and *A. steueri.* I. Some environmental factors affecting egg development and the nature of resting eggs, *Bull. Plankton Soc. Jpn.,* 27, 1, 1980.

60. McLaren, I. A., Predicting development rate of copepod eggs, *Biol. Bull.,* 131, 457, 1966.

61. McLaren, I. A., Corkett, C. J., and Zillioux, E. J., Temperature adaptations of copepod eggs from the arctic to the tropics, *Biol. Bull.,* 137, 486, 1969.

62. Bĕlehrádek, J., *Temperature and Living Matter, Protoplasma Monographien,* Vol. 8, 1, Borntraeger, Berlin, 1935.

63. McLaren, I. A., Walker, D. A., and Corkett, C. J., Effects of salinity on mortality and development rate of eggs of the copepod *Pseudocalanus minutus, Can. J. Zool.,* 46, 1267, 1968.

64. Uye, S. and Fleminger, A., Effect of various environmental factors on egg development of several species of *Acartia* in Southern California, *Mar. Biol.,* 38, 253, 1976.

65. Heinle, D. R., Production of a calanoid copepod, *Acartia tonsa,* in the Patuxent River Estuary, *Chesapeake Sci.,* 7, 59, 1966.

66. Miller, C. B., Johnson, J. K., and Heinle, D. R., Growth rules in the marine copepod genus *Acartia, Limnol. Oceanogr.,* 22, 326, 1977.

67. Mullin, M. M. and Brooks, E. R., Growth and metabolism of two planktonic, marine copepods as influenced by temperature and type of food, in *Marine Food Chains,* Steele, J. H., Ed., University of California Press, Berkeley, 1970, 74.

68. Uye, S., Population dynamics and production of *Acartia clausi* Giesbrecht (Copepoda: Calanoida) in inlet waters, *J. Exp. Mar. Biol. Ecol.,* 57, 55, 1982.

69. Grice, G. D. and Marcus, N. H., Dormant eggs of marine copepods, *Oceanogr. Mar. Biol. Ann. Rev.,* 19, 125, 1981.

70. Sazhina, L. I., Hibernating eggs of marine Calanoida, *Zool. Zh.,* 47, 1554, 1968. (Fish. Res. Bd. Canada, Transl. Ser., 1259, 1).

71. Uye, S., Kasahara, S., and Onbé, T., Calanoid copepod eggs in sea-bottom muds. IV. Effects of some environmental factors on the hatching of resting eggs, *Mar. Biol.,* 51, 151, 1979.

72. Kasahara, S., Uye, S., and Onbé, T., Calanoid copepod eggs in sea-bottom muds, *Mar. Biol.,* 26, 167, 1974.

73. Marcus, N. H., Photoperiodic control of diapause in the marine calanoid copepod *Labidocera aestiva, Biol. Bull.,* 159, 311, 1980.

74. Kasahara, S. and Uye, S., Calanoid copepod eggs in sea-bottom muds. V. Seasonal changes in hatching of subitaneous and diapause eggs of *Tortanus forcipatus, Mar. Biol.,* 55, 63, 1979.

75. Brewer, R. H., The phenology of *Diaptomus stagnalis* (Copepoda: Calanoida): the development and the hatching of the egg stage, *Physiol. Zool.,* 37, 1, 1964.

76. Elgmork, K., Ecological aspects of diapause in copepods, *Proc. Symp. Crustacea,* 3, 947, 1967.

77. Elgmork, K., Bottom resting stages of planktonic cyclopoid copepods in meromictic lakes, *Verh. Int. Verein. Limnol.,* 18, 1474, 1973.

78. Spindler, K.-D., Dormanzauslosung und Dormanzcharakteristika beim Süsswassercopepoden *Cyclops vicinus, Zool. Jahrb. Abt. Physiol.,* 76, 139, 1971.

79. Nilssen, J. P. and Elgmork, K., *Cyclops abyssorum* — life cycle dynamics and habitat selection, *Mem. Ist. Ital. Idrobiol.,* 34, 197, 1977.

80. Elgmork, K., Dynamics of zooplankton communities in some small inundated ponds, *Folia Limnol. Scand.,* 12, 1, 1964.

81. Landry, M. R., Dark inhibition of egg hatching of the marine copepod *Acartia clausi* Giesbr., *J. Exp. Mar. Biol. Ecol.,* 20, 43, 1975.

82. Coull, B. C. and Dudley, B. W., Delayed naupliar development of meiobenthic copepods, *Biol. Bull.,* 150, 38, 1976.

83. Corner, E. D. S., Head, R. N., Kilvington, C. C., and Marshall, S. M., On the nutrition and metabolism of zooplankton. IX. Studies relating to the nutrition of overwintering *Calanus, J. Mar. Biol. Assoc. U.K.,* 54, 319, 1974.

84. Grainger, E. H., The annual oceanographic cycle at Igloolik in the Canadian Arctic. I. The zooplankton and physical and chemical observations, *J. Fish. Res. Board Can.,* 16, 453, 1959.

85. Ussing, H. H., The biology of some important plankton animals in the fjords of East Greenland, *Medd. Grønl.,* 100, 1, 1938.

86. Jespersen, P., Investigations on the copepod fauna in East Greenland, *Medd. Grønl.,* 119, 1, 1939.

87. Østvedt, O.-J., Zooplankton investigations from Weather Ship M in the Norwegian Sea, *Hvalradets Skr.,* 40, 1, 1955.

88. McLaren, I. A., Population and production ecology of zooplankton in Ogac Lake, a landlocked fjord on Baffin Island, *J. Fish. Res. Board Can.,* 26, 1485, 1969.

89. Carter, J. C. H., The ecology of the calanoid copepod *Pseudocalanus minutus* Kroyer in Tessiarsuk, a coastal meromictic lake of northern Labrador, *Limnol. Oceanogr.,* 10, 345, 1965.

90. Runnstrøm, S., Eine Uebersicht uber das Zooplankton des Herdla- und Hjeltefjordes, *Bergens Mus. Årb. Naturvid, Rekke,* 1931, 1, 1932.

91. Coull, B. C. and Grant, J., Encystment discovered in a marine copepod, *Science,* 212, 342, 1981.

92. Mullin, M. M. and Brooks, E. R., Laboratory cultures, growth rate, and feeding behavior of a planktonic marine copepod, *Limnol. Oceanogr.,* 12, 657, 1967.

93. Bernard, M., Premières observations sur la ponte, les oeufs, les stades naupliens et l'alimentation de quatre Pontellides (Copepoda), *Rapp. P.-V. Réun. Comm. Int. Explor. Sci. Mér Meriterr. Monaco,* 19, 525, 1968.

94. Conover, R. J., Reproductive cycle, early development, and fecundity in laboratory populations of the copepod *Calanus hyperboreus, Crustaceana,* 13, 61, 1967.

95. Marshall, S. M. and Orr, A. P., On the biology of *Calanus finmarchicus.* IX. Feeding and digestion in the young stages, *J. Mar. Biol. Assoc. U.K.,* 35, 587, 1956.

96. Gauld, D. T., Swimming and feeding in crustacean larvae: the nauplius larva, *Proc. Zool. Soc. Lond.,* 132, 31, 1959.

97. Lewis, A. G., An enrichment solution for culturing the early development stages of the planktonic marine copepod *Euchaeta japonica* Marukawa, *Limnol. Oceanogr.,* 12, 147, 1967.

98. Nicholls, A. G., The developmental stages of *Euchaeta norvegica* Boeck, *Proc. R. Soc. Edinburgh,* Sect. B, 54, 31, 1934.

99. Bernard, M., Le développement nauplien de deux Copépodes carnivores: *Euchaeta marina* (Prestandr.) et *Candacia armata* (Boeck) et observations sur le cycle de l'astaxanthine au cours de l'ontogénèse, *Pelagos,* 2, 51, 1964.

100. Sekiguchi, H., (The origin and adaptation of ontogenetic vertical migrations of the pelagic zooplankton in the sea. 1.), *Kaiyo Kagaku,* 8, 63 (in Japanese), *Can. Fish. Mar. Serv., Transl. Ser.,* 4134, 1, 1977.

101. Benson, A. A. and Lee, R. F., The role of wax in oceanic food chains, *Sci. Am.,* 232, 76, 1975.

102. Storch, O., Der Nahrungserwerb zweier Copepodennauplien (*Diaptomus gracilis* und *Cyclops strenuus*), *Zool. Jahrb., Abt. Allg. Zool. Physiol. Tiere,* 45, 385, 1928.

103. Paffenhöfer, G.-A., Grazing and ingestion rates of nauplii, copepodids and adults of the marine planktonic copepod *Calanus helgolandicus, Mar. Biol.,* 11, 286, 1971.

104. Corkett, C. J. and McLaren, I. A., Relationships between development rate of eggs and older stages of copepods, *J. Mar. Biol. Assoc. U.K.,* 50, 161, 1970.

105. Paffenhöfer, G.-A. and Harris, R. P., Feeding, growth and reproduction of the marine copepod *Pseudocalanus elongatus* Boeck, *J. Mar. Biol. Assoc. U.K.,* 56, 327, 1976.

106. Petipa, T. S., Oxygen consumption and food requirement in the copepods *Acartia clausi* Giesbr. and *A. latisetosa* Kritcz., *Zool. Zhurn.,* 45, 363, (*Fish. Lab. Lowestoft, Transl. n. s.,* 90, 1, 1966).

107. Nassogne, A., Influence of food organisms on the development and culture of pelagic copepods, *Helgolander wiss. Meeresunters.,* 20, 333, 1970.

108. Lewis, A. G. and Ramnarine, A., Some chemical factors affecting the early developmental stages of *Euchaeta japonica* (Crustacea: Copepoda: Calanoida) in the laboratory, *J. Fish. Res. Board Can.*, 26, 1347, 1969.

109. Esterley, C. O., The feeding habits and food of pelagic copepods and the question of nutrition by organic substances in solution in the water, *Univ. Calif. Publ. Zool.*, 16, 171, 1916.

110. Alcaraz, M., Paffenhöfer, G.-A., and Strickler, J. R., Catching the algae: a first account of visual observations on filter-feeding calanoids, in *Evolution and Ecology of Zooplankton Communities*, Kerfoot, W. C., Ed., 1980, 241.

111. Koehl, M. A. R. and Strickler, J. R., Copepod feeding currents: food capture at low Reynolds number, *Limnol. Oceanogr.*, 26, 1062, 1981.

112. Price, H. J., Paffenhöfer, G.-A., and Strickler, J. R., Modes of cell capture in calanoid copepods, *Limnol. Oceanogr.*, 28, 116, 1983.

113. Cowles, T. J. and Strickler, J. R., Characterization of feeding activity patterns in the planktonic copepod *Centropages typicus* Kröyer under various food conditions, *Limnol. Oceanogr.*, 28, 106, 1983.

114. Beklemishev, K. V., Obnaruzhenie kremnevych obrazovanii v kozhnych pokrovach nizshich raboobraznich (The finding of silica formations in the integuments of Entomostraca), *Dokl. Akad. Nauk SSSR* (in Russian), 97, 543, 1954.

115. Beklemishev, K. V., (On the anatomy of masticatory organs of Copepoda. 2. The masticatory edge in mandibles of certain species of Calanidae and Eucalanidae), *Tr. Inst. Okeanol. Akad. Nauk SSSR* (in Russian), 30, 148, 1959.

116. Sullivan, B. K., Miller, C. B., Peterson, W. T., and Soeldmer, A. H., A scanning electron microscope study of the mandibular morphology of boreal copepods, *Mar. Biol.*, 30, 175, 1975.

117. Conover, R. J., Oceanography of Long Island Sound, 1952—1954. VI. Biology of *Acartia clausi* and *A. tonsa.*, *Bull. Bingham Oceanogr. Collect. Yale Univ.*, 15, 156, 1956.

118. Itoh, K., A consideration on feeding habits of planktonic copepods in relation to the structure of their oral parts, *Bull. Plankton Soc. Jpn.*, 17, 1, 1970.

119. Anraku, M. and Omori, M., Preliminary survey of the relationship between the feeding habit and the structure of the mouth-parts of marine copepods, *Limnol. Oceanogr.*, 8, 116, 1963.

120. Turner, J. T., Scanning electron microscope investigations of feeding habits and mouthpart structures of copepods of the family Pontellidae, *Bull. Mar. Sci.*, 28, 487, 1978.

121. Sars, G. O., *Crustacea of Norway*, Vol. 4, Copepoda Calanoida, Bergen Museum, Bergen, 1, 1903.

122. Wickstead, J. H., Food and feeding in pelagic copepods, *Proc. Zool. Soc. Lond.*, 139, 545, 1962.

123. Nival, P. and Nival, S., Efficacité de filtration des Copépodes planctoniques, *Ann. Inst. Océanogr.*, 49, 135, 1973.

124. Nival, P. and Nival, S., Particle retention efficiencies of an herbivorous copepod, *Acartia clausi* (adult and copepodite stages): effects on grazing, *Limnol. Oceanogr.*, 21, 24, 1976.

125. Poulet, S. A., Grazing of *Pseudocalanus minutus* on naturally occurring particulate matter, *Limnol. Oceanogr.*, 18, 564, 1973.

126. Poulet, S. A., Seasonal grazing of *Pseudocalanus minutus* on particles, *Mar. Biol.*, 25, 109, 1974.

127. Poulet, S. A., Comparison between five coexisting species of marine copepods feeding on naturally occurring particulate matter, *Limnol. Oceanogr.*, 23, 1126, 1978.

128. Wilson, D. S., Food size selection among copepods, *Ecology*, 54, 909, 1973.

129. Robertson, S. B. and Frost, B. W., Feeding by an omnivorous planktonic copepod, *Aetidius divergens* Bradford, *J. Exp. Mar. Biol. Ecol.*, 29, 231, 1977.

130. Frost, B. W., Effects of size and concentration of food particles on the feeding behavior of the marine planktonic copepod *Calanus pacificus*, *Limnol. Oceanogr.*, 17, 805, 1972.

131. Huntley, M., Nonselective, nonsaturated feeding by three calanoid copepod species in the Labrador Sea, *Limnol. Oceanogr.*, 26, 831, 1981.

132. Heinle, D. R., Harris, R. P., Ustach, J. F., and Flemer, D. A., Detritus as food for estuarine copepods, *Mar. Biol.*, 40, 341, 1977.

133. Harding, G. C. H., The food of deep-sea copepods, *J. Mar. Biol. Assoc. U.K.*, 54, 141, 1974.

134. Arashkevich, Ye. G., (The food and feeding of copepods in the northwestern Pacific), *Okeanologiya*, 9, 857, Engl. transl., *Oceanology*, 9, 695, 1969.

135. Arashkevich, Ye. G., (Vertical distribution of trophic groups of copepods in the boreal and tropical regions of the Pacific Ocean), *Okeanologiya*, 12, 315, Engl. transl., *Oceanology*, 12, 265, 1972.

136. Lampitt, R. S., Carnivorous feeding by a small marine copepod, *Limnol. Oceanogr.*, 23, 1228, 1978.

137. Mullin, M. M., Differential predation by the carnivorous marine copepod, *Tortanus discaudatus*, *Limnol. Oceanogr.*, 24, 774, 1979.

138. Lonsdale, D. J., Heinle, D. R., and Siegfried, C., Carnivorous feeding behavior of the adult calanoid copepod *Acartia tonsa*, *J. Exp. Mar. Biol. Ecol.*, 36, 235, 1979.

139. Landry, M. R., Predatory feeding behavior of a marine copepod, *Labidocera trispinosa, Limnol. Oceanogr.*, 23, 1103, 1978.

140. Zaret, T. M., *Predation and Freshwater Communities,* Yale Univ. Press, New Haven, 1, 1980.

141. Båmstedt, U. and Holt, M. R., Experimental studies on the deep-water pelagic community of Korsfjorden, Western Norway. Prey size preference and feeding of *Euchaeta norvegica (Copepoda), Sarsia,* 63, 225, 1978.

142. Lillelund, K. and Lasker, R., Laboratory studies of predation by marine copepods on fish larvae, *Fish. Bull.,* 69, 655, 1971.

143. Wickstead, J., A predatory copepod, *J. Anim. Ecol.,* 28, 69, 1959.

144. Davis, C. C., *Sagitta* as food for *Acartia, Astarte,* 10, 1, 1977.

145. Alvarez, V. and Matthews, J. B. L., Experimental studies on the deep-water pelagic community of Korsfjorden, Western Norway. Feeding and assimilation by *Chiridius armatus* (Crustacea, Copepoda), *Sarsia,* 58, 67, 1975.

146. Paffenhöfer, G.-A. and Knowles, S. C., Feeding of marine planktonic copepods on mixed phytoplankton, *Mar. Biol.,* 48, 143, 1978.

147. Poulet, S. A. and Marsot, P., Chemosensory grazing by marine calanoid copepods (Arthropoda: Crustacea), *Science,* 200, 1403, 1978.

148. Strickler, J. R. and Bal, A. K., Setae of the first antennae of the copepod *Cyclops scutifer* (Sars): their structure and importance, *Proc. Natl. Acad. Sci. USA,* 70, 2656, 1973.

149. Strickler, J. R., Intra- and interspecific information flow among planktonic copepods: receptors, *Verh. Internat. Verein. Limnol.,* 19, 1951, 1975.

150. Gerritsen, J., Intra-specific swimming patterns and predation of planktonic copepods, *Verh. Internat. Verein. Limnol.,* 20, 2531, 1978.

151. Cannon, H. G., On the feeding mechanism of the copepods, *Calanus finmarchicus* and *Diaptomus gracilis, Br. J. Exper. Biol.,* 6, 131, 1929.

152. Lowndes, A. G., The swimming and feeding of certain calanoid copepods, *Proc. Zool. Soc. Lond.,* 1935, 687, 1935.

153. Gauld, D. T., The swimming and feeding of planktonic copepods, in *Some Contemporary Studies in Marine Science,* Barnes, H., Ed., Hafner, Darien, Conn., 1966, 313.

154. Turner, J. T., Zooplankton feeding ecology: Contents of fecal pellets of the copepods *Eucalanus pileatus* and *Paracalanus quasimodo* from continental shelf waters of the Gulf of Mexico, *Mar. Ecol. Prog. Ser.,* 15, 27, 1984.

155. Pennel, W. M., Description of a new species of pontellid copepod, *Anomalocera opalus,* from the Gulf of St. Lawrence and shelf waters of the northwest Atlantic Ocean, *Can. J. Zool.,* 54, 1664, 1976.

156. Hempel, G. and Weikert, H., The neuston of the tropical and boreal northeastern Atlantic Ocean, *Mar. Biol.,* 13, 70, 1972.

157. Zaitsev, Yu. P., *Marine Neustonology,* Israel Program Sci. Transl., Jerusalem, 1, 1971.

158. Longhurst, A. R., Vertical migration, in *The Ecology of the Seas,* Cushing, D. H. and Walsh, J. J., Eds., Blackwell Scientific Publications, London, 1976, chap. 6.

159. Eriksson, S., The biology of marine planktonic Copepoda on the west coast of Sweden, *Zoon,* 1, 37, 1973.

160. Nicholls, A. G., On the biology of *Calanus finmarchicus.* III. Vertical distribution and diurnal migration in the Clyde Sea-area, *J. Mar. Biol. Assoc. U.K.,* 19, 139, 1933.

161. Russell, F. S., The vertical distribution of plankton in the sea, *Biol. Rev.,* 2, 213, 1927.

162. Kikuchi, K., Diurnal migration of plankton Crustacea, *Q. Rev. Biol.,* 5, 189, 1930.

163. Cushing, D. H., The vertical migration of planktonic Crustacea, *Biol. Rev.,* 26, 158, 1951.

164. Bainbridge, R., Migrations, in *The Physiology of Crustacea,* Vol. 2, Waterman, T. H., Ed., Academic Press, New York, 1961, 431.

165. Banse, K., On the vertical distribution of zooplankton in the sea, *Prog. Oceanogr.,* 2, 53, 1964.

166. Vinogradov, M. E., *Vertical Distribution of the Oceanic Zooplankton,* Israel Program Sci. Transl., Jerusalem, 1, 1970.

167. Bohrer, R. N., Experimental studies on diel vertical migration, in *Evolution and Ecology of Zooplankton Communities,* Kerfoot, W. C., Ed., University Press, Hanover, 1980, 111.

168. Marshall, S. M. and Orr, A. P., On the biology of *Calanus finmarchicus.* XI. Observations on vertical migration especially in female *Calanus, J. Mar. Biol. Assoc. U.K.,* 39, 135, 1960.

169. Hayward, T. L., Mating and the depth distribution of an oceanic copepod, *Limnol. Oceanogr.,* 26, 374, 1981.

170. Davis, C. C., *The Marine and Fresh-water Plankton,* Michigan State Univ. Press, East Lansing, 1, 1955.

171. Zaret, T. M. and Suffern, J. S., Vertical migration in zooplankton as a predator avoidance mechanism, *Limnol. Oceanogr.,* 21, 804, 1976.

172. McLaren, I. A., Effects of temperature on growth of zooplankton and the adaptive value of vertical migration, *J. Fish. Res. Board Can.*, 20, 685, 1963.
173. Enright, J. T., Diurnal vertical migration: adaptive significance and timing: a metabolic model, *Limnol. Oceanogr.*, 22, 856, 1977.
174. Enright, J. T. and Honegger, H.-W., Diurnal vertical migration: adaptive significance and timing. 2. Test of the model: details of timing, *Limnol. Oceanogr.*, 22, 873, 1977.
175. Herman, S., Vertical migration of the opossum shrimp, *Neomysis americana* Smith, *Limnol. Oceanogr.*, 8, 228, 1963.
176. Dingle, H., The occurrence and ecological significance of color responses in some marine crustaceans, *Am. Nat.*, 96, 151, 1962.
177. Itoh, K., Studies on the vertical migration of zooplankton in relation to the conditions of underwater illumination, *Sci. Bull. Fac. Agric. Kyushu Univ.*, 25, 71, 1970 (Japanese, Engl. summ.).
178. Buchanan, C. and Haney, J. F., Vertical migrations of zooplankton in the Arctic: A test of the environmental controls, in *Evolution and Ecology of Zooplankton Communities,* Kerfoot, W. C., Ed., University Press, Hanover, 1980, 69.
179. Bogorov, V. G., Peculiarities of diurnal vertical migrations of zooplankton in polar seas, *J. Mar. Res.*, 6, 25, 1946.
180. Russell, F. S., The vertical distribution of marine macroplankton. IV. The apparent importance of light intensity as a controlling factor in the behaviour of certain species in the Plymouth area, *J. Mar. Biol. Assoc. U.K.*, 14, 415, 1926.
181. Strickler, J. R., *Experimentell-oekologische Untersuchungen über die Vertikalwanderung planktischer Crustaceen,* Doctoral thesis, Eidgenössischen Technischen Hochschule, Zurich, 1, 1969.
182. Bainbridge, R., Studies on the interrelationships of zooplankton and phytoplankton, *J. Mar. Biol. Assoc. U.K.*, 32, 385, 1953.

Chapter 5

ASPECTS OF THE PHYSIOLOGY AND ECOLOGY OF PELAGIC LARVAE OF MARINE BENTHIC INVERTEBRATES

Randy Day and Larry McEdward

TABLE OF CONTENTS

I. INTRODUCTION

Pelagic larvae of marine benthic invertebrates comprise the meroplanktonic component of the zooplankton community. Most major taxa of benthic invertebrates possess some species with pelagic larvae. In light of the diversity of larval forms, an overview of classification schemes for patterns of larval development will serve as an introduction. The emphasis of the paper will be on three topics of general importance in complex animal life cycles: (1) causes of mortality during the planktonic period, (2) energetics of larval development and (3) dispersal. Energetics and dispersal are central to most recent analyses of the adaptive nature of various patterns of larval development. By concentrating on general features of pelagic larval stages, we hope to illustrate some important differences between meroplanktonic and holoplanktonic animals.

A larva can be defined as the developmental stage(s) between hatching from the egg membranes and metamorphosis to the adult form.[1] Typically such stages possess "larval structures" that exist or function only during that period of the life cycle. The diversity of larval forms and the variety of complex life cycles present throughout marine benthic invertebrate classes cannot be adequately captured in any single, concise definition. The present intent is to indicate the phase of the life cycle that is planktonic.

The terms pelagic and planktonic (sensu Hardy) will be used interchangeably with reference to invertebrate larvae.[2]

A. Classification of Larval Patterns

Thorson classified larvae of marine bottom invertebrates into three basic types:[3] (1) larvae that develop in the plankton (pelagic); (2) larvae that develop in an egg mass completely omitting a pelagic phase (direct); and (3) larvae that develop within the parental organism. He suggested that the pelagic type of development be divided into two subtypes: (a) lecithotrophic development, where larvae do not feed during planktonic life; and (b) planktotrophic development where larvae actively feed on planktonic organisms.

Thorson's classification has been extended to include: (1) pelagic development divided into subtypes, lecithotrophy or planktotrophy; (2) demersal development; (3) direct development; and (4) viviparity.[1,4-7] Energy considerations, such as partitioning of reproductive effort, have been the basis of recent models comparing different patterns of larval development. Vance defined reproductive efficiency as the ratio of larvae surviving to metamorphosis per unit of energy expended in reproduction.[8,9] Under the assumptions of a simplified model, lecithotrophic and planktotrophic development were shown to be extremes maximizing reproductive efficiency, intermediate modes of development were found to be less efficient. Underwood criticized Vance's models on the grounds that there was no correlation between developmental period and egg size.[8,10] However, a positive correlation between developmental period and egg size has been shown in a large variety of marine organisms.[11]

Christensen and Fenchel, in 1979, predicted that only the extremes of the possible range of egg size and methods of nutrition (i.e., planktotrophy and lecithotrophy) are evolutionarily stable.[12] They also concluded that over a certain range of environmental parameters, the two developmental modes are both evolutionarily stable, with planktotrophy more efficient than lecithotrophy when planktonic food is abundant and planktonic predation is low, and with lecithotrophy more efficient when either or both of these conditions are reversed.

These hypotheses suggest that at least two options are open to a species of benthic invertebrate. An organism may concentrate sufficient energy within each larva to enable it to be nonfeeding and nutritionally independent of its surroundings during the larval stage, or it may produce feeding larvae which accumulate the energy needed for larval development and metamorphosis during pelagic life. Species adopting the former strategy must produce larger eggs rich in yolk. Species which adopt the latter strategy can afford to produce many small, yolk-poor eggs, but their larvae will be at a greater risk from predation, starvation, or excessive dispersal because of the longer planktonic period.[5]

The evolutionary advantage of each strategy is not well understood. For example, Chia maintained that lecithotrophic development is cheaper in terms of energy cost of reproduction; however, it is not as evolutionarily advantageous as planktotrophy.[1] Todd and Doyle on the other hand, believe that lecithotrophy is energetically more expensive than planktotrophy in terms of caloric output by the adult.[13] Menge considered brooding (lecithotrophy) in a starfish to be the consequence of competitor induced small size, suggesting that a small species that broadcasts larvae would not produce enough offspring to replace itself.[14] This conclusion is based on the fact that the seastar *Leptasterias hexactis*, which broods its few annual offspring, has a much higher relative investment in its gonads than does the related species *Pisaster ochraceas* which produces millions of planktonic larvae per season.

Todd provided evidence to suggest that small organisms with little energy to devote to reproduction are limited to planktotrophy.[13,15-17] Todd stated that this concept cannot describe the adaptive nature of these reproductive patterns because reproductive effort seems not be a primary determinant of the developmental type or the number of eggs produced per individual.[15] Strathmann has suggested that egg size may result from selection for increased larval size or increased size at settling rather than selection for a specific reproductive strategy.[18] Some workers doubt that evolution of the mode of reproduction is governed by energy considerations at all.[10,18,19,20]

It is apparent that there remains a great deal of doubt as to the ecological or evolutionary mechanisms responsible for the diversity of reproductive strategies. However, some interesting relationsips exist. For example, morphological evidence suggests that species with nonfeeding larvae have evolved from species with feeding larvae more frequently than the reverse, and that this has occurred numerous times in many classes of marine invertebrates.[3,21-23] This is not surprising because the change to lecithotrophy often involves loss of elaborate larval feeding structures. Evolution of a new set of larval feeding structures would require a more complex series of changes. Thus, once a species has evolved a nonfeeding larva, planktotrophy is no longer an option.[23]

Eyster noted that although developmental modes often are considered species specific and inflexible, the data suggest otherwise. Modes may differ between geographically separated populations,[22,24] or within a single population via twinning,[22] via nurse egg consumption,[25] or in accordance with environmental conditions.[26] Pelagic and nonpelagic development occurs within a single population of the nudibranch *Tenellia pallida* from South Carolina, where both developmental types occur simultaneously, under identical conditions, and without the aid of nurse eggs.[27]

In spite of this apparent flexibility in developmental modes, approximately 70% of the species of marine bottom invertebrates reproduce by means of pelagic planktotrophic development.[22] This suggests that species with planktotrophic larvae possess, at least occasionally, some evolutionary or ecological advantages compared to species possessing the other types of development.

II. THE PELAGIC PERIOD

A. Duration

The pelagic period in the life cycle of benthic marine invertebrates typically begins with spawning or hatching. Free spawning of gametes by both sexes and ectosomatic fertilization (e.g., echinoderms) results in pelagic zygotes, embryos, and larvae. Some taxa release pelagic larvae following benthic development of embryos in capsules (e.g., some mollusks and some polychaetes) or in the adult body (e.g., colonial ascidians). Initiation of swimming occurs subsequent to hatching in all of these patterns.

The duration of the pelagic period ranges from minutes to years. The polychaete, *Spirorbis borealis*, has a planktonic life of 30 minutes to several hours.[28] The tadpole larva of the compound ascidian, *Trididemnum solidum*, swims for only 2 to 15 minutes.[29] Typical maximum pelagic periods range up to approximately 2 years (e.g., 22 months for the phyllosoma larva of the lobster, *Janus edwardsii*).[30]

Traditionally, pelagic and nonpelagic development have been contrasted as fundamentally different larval patterns.[22] These short pelagic periods suggest that a continuum exists from entirely benthic development to long planktonic larval periods. The difference in scale of dispersal, degree of larval adaptation to planktonic life, and consequence for recruitment may be as great for species with extremely short and extremely long pelagic phases as are the differences between a very short larval period and entirely benthic development.

Typical pelagic periods for lecithotrophic larvae are in the range of several days to 2 weeks. Planktotrophic larvae typically require 2 to 6 weeks to develop to a metamorphically competent state. These estimates generally result from observations of laboratory cultures. Pelagic periods have rarely been measured in the field with accuracy. However, Jorgensen followed a cohort of the bivalve mollusk, *Mytilus edulis*, and obtained an estimate of pelagic duration of one month.[31]

Field studies are difficult because of the problem of sampling an identifiable cohort of larvae, the lack of knowledge of larval behavior, the potential for asynchronous spawning, the potential for delayed settlement and the tremendous variation in development times within a cohort. For example, the heart urchin, *Brissaster latifrons*, in a single laboratory culture, had a minimum larval life of 9 weeks and a maximum of 24 weeks.[32]

In general, the pelagic phase is short, relative to the benthic juvenile and adult life span. However, it can be of considerable importance in the life cycle. Many species (e.g., sessile adults) rely entirely on the larval stages for dispersal.

B. Sources of Mortality

It is obvious from the difference between fecundity and recruitment that there is tremendous mortality during the larval period.[22] Many benthic free-spawners produce 100,000 to 1,000,000 eggs per spawning female, yet benthic populations recruit but a small fraction of these individuals. Menge estimated the fecundity of the starfish, *Pisaster ochraceus*, (400g) at 40 million eggs per female each year and by assuming that the benthic population size was at equilibrium, calculated a larval and early juvenile mortality of 0.99999999854.[14]

Very little is known about sources of larval mortality in nature. Predation, physical environmental factors, inappropriate scales of dispersal and starvation have been suggested.[5,22] Certainly, physical environmental factors (e.g., temperature and salinity) often exceed larval tolerances. Laboratory studies and field observations have demonstrated that the reproductive season and geographic range often coincide with the embryonic and larval tolerances of a species.[22,33,34]

Thorson, and many others since, have argued explicitly or have assumed implicitly that lethal physical conditions are quantitatively unimportant as a direct cause of planktonic larval mortality.[22] However, sublethal stress may have important consequences through retardation or acceleration of rates of development.[35-44] There are several possible effects: (1) rate changes during periods of active morphogenesis could disrupt finely coordinated events;[45] (2) changes in metabolic rates and feeding rates could have important long-term effects on the maintenance of energy balances.[46] These two phenomena have received very little attention to date. A third possible effect of sublethal stress is simply a change in the duration of the planktonic period, thus increasing the risk from other sources of mortality.

Although sublethal stress typically reduces survival and prolongs development,[37] conditions for maximal larval survival do not always yield rapid development. Akesson measured growth rates of larvae, juveniles, and adults, as well as adult survival and reproductive output of the polychaete, *Ophryotrocha labronica*, as a function of temperature.[47] High temperature (28°C) resulted in the highest larval, juvenile, and adult mortality and the lowest fecundity. However, the short generation time at 28°C more than compensated for these losses, and resulted in the greatest rate of population increase. This result suggests that caution be exercised when interpreting environmental conditions as optimal or suboptimal from studies of their effects on a single life history stage.

The potential duration of the pelagic period has a positive correlation with the scale of dispersal from the parent. A significant change in the rate of development may increase the chance of being swept beyond the appropriate adult habitat (e.g.,[48] over ocean basins). Alternatively, it may result in insufficient spread to encounter appropriate settlement sites (e.g., opportunistic species with emphemeral habitats). The appropriate scale of dispersal is intimately associated with the ecology of the benthic stages and it is not possible to generalize from a given change in pelagic duration to the expected consequences.

There seem to be different degrees of response by different species to factors that perturb developmental rates. Many species of crustaceans have a typical sequence and a number of larval instars. Temperature, salinity or nutritional stress can increase or decrease the number and duration of the larval stages.[49,50] But species in which there is a premium on larval behavior for recruitment to the parental population (e.g., *Rhithropanopeus harrisii*) are relatively invariant in molt sequence under sublethal experimental treatments, thereby minimizing changes in the pelagic period.[44] As a consequence, it is difficult to make general predictions about the effect of sublethal environmental stress on development rate and dispersal potential.

Predation has been considered the single most important cause of larval loss.[22] A correlation exists between fecundity and the duration of the larval period but the causal basis is unclear. It is known that larvae are subject to predation;[22,51] for example, planktivorous fish eat lecithotrophic starfish embryos,[52] estuarine ctenophores are known to be predators of many larval forms, and benthic suspension feeders eat eggs and larvae in the laboratory.[53,54] The quantitative effect of such predation in nature is unknown. Indirect evidence for the importance of predation comes from defensive features in larvae. Defense against predation has been suggested as a function of the rostral spines of zoea larvae,[53] the serrated setae of polychaete larvae, and for torsion in gastropod veligers.[55,56] In one case, chemical defenses (saponins) have been demonstrated to be effective against fish predation.[52] Larval behavior can be effective in avoiding predators. If benthic suspension feeders are voracious larvivores, then positive phototaxis early after hatching could move larvae up into the water column away from predators.[54]

A well-documented case in which larval behavior has been demonstrated to be effective against a predator involves the xanthid crab, *Rhithropanopeus harrisii*. A shadow reflex of sinking or active downward swimming occurs in all of the zoeal stages in response to a sudden decrease in light intensity.[57] The response is superimposed over any other swimming response and has been shown to be an effective means of avoiding the ctenophore, *Mnemiopsis leidyi*, in the laboratory.[58] Chemical cues are not required to elicit the response. Field studies suggest that the shadow response is effective in nature. Gut contents show that a large number of larval forms are eaten by ctenophores in the field. But zoeae of *R. harrisii* are not among the prey.[53] Elimination of the shadow response, through experimental modification of the illumination, leads to ingestion of large numbers of zoeae by ctenophores in the lab.[58] A similar response occurs in at least six other families of estuarine crabs.[59]

Acceptance of the assumption that predation is the most important source of larval mortality has led to a series of models of marine life histories.[5,8,9,12,60,61] A common feature of these models is the assumption of a constant and significant mortality rate. Not surprisingly, these models show favorable consequences for traits and conditions that combine to minimize exposure to predation (i.e., minimal planktonic duration). From such a perspective, sublethal environmental stresses that alter the rate of development and therefore the duration of the pelagic period assume additional importance.

Starvation is the most controversial, if least understood, cause of larval mortality. Lecithotrophic larvae, or planktotrophic larvae during a lecithotrophic period (e.g., prefeeding embryos or terminal settlement stages), are independent of planktonic food conditions. The possibility of starvation should not exist. However, specific requirements for a substratum at settlement may involve a significant delay in metamorphosis.[4,62] Energetic considerations for an organism living on a fixed amount of fuel are not trivial.[4,63,64] Any factor that changes the rate of development, the rate of metabolism, or the time required to encounter appropriate substrata for settling could produce the energetic equivalent of starvation.

Is starvation an important source of mortality in planktotrophic larvae? At present the data are insufficient to yield reliable generalizations. Two schools of thought exist. First, starvation resistance in the lab is quite high in some species, and signs of starvation are never observed in the field.[3,22,65,66] These observations, combined with a lack of knowledge of larval nutritional requirements, have led some authors to argue that starvation is unimportant as a cause of larval loss.[8,9,12,22,66] Second, laboratory studies on other larval forms (primarily decapod zoeae) have shown susceptibility to starvation even at food particle concentrations several fold higher than typical concentrations in nature.[67-70] These data suggest that starvation is a significant source of mortality for all larvae except those that encounter rich patches of plankton.

Associated with starvation is the notion of food limitation. At present it is not possible to generalize as to whether or not planktotrophic larval growth rates are retarded by natural food concentrations. Development rates can be limited by food concentration in laboratory culture.[71-73] Qualitative nutritional requirements can also influence larval growth and survival.[72-75] If food limitation occurs it may act in concert with other sources of mortality (e.g., temperature) to produce a very complicated set of determinants of larval mortality.[75]

Regardless of causes, much mortality occurs during the larval phase. A fundamental, yet unresolved, question is: to what extent is larval mortality genotype specific? If mortality is primarily selective, then powerful selection pressure is exerted on the larval phase of the life cycle, and evolution of larval traits should be limited by the rate of appearance of favorable mutations; fixation should be common. Alternatively, if larval death is random with respect to genotype, then considerable genetic drift occurs and larval features will drift toward fixation without regard to the adaptiveness of such features. Such a situation would greatly complicate our efforts to understand the ecological and evolutionary determinants of larval patterns. At present, too little is known about the adaptiveness of larval features, sources of mortality, and phenotypic variation to decide among these two alternatives and the myriad intermediate possibilities.

Table 1

MAIN BIOCHEMICAL CONSTITUENTS OF SOME MARINE
INVERTEBRATE LARVAE[78]

Species	Larval type	Mean dry wt. (μg)	Lipid (%)	Carbohydrate (%)	Protein (%)
Mollusca					
Ostrea edulis	Veliger	1.0	6.6	1.5	48.1
Pecten maximus	Veliger	0.1	7.9	1.4	—
Littorina littorea	Veliger	0.9	10.3	1.8	20.0
Littorina neritoides	Veliger	1.0	9.1	0.9	22.5
Dendropoma corallinaceum	Veliger	73.0	12.2	1.9	36.2
Crustacea					
Eliminius modestus	Nauplius	0.4	8.1	2.0	63.7
Sacculina carcini	Nauplius	1.0	16.5	3.3	—
Balanus hameri	Nauplius	3.1	16.6	2.5	55.0
Balanus balanoides	Nauplius	1.1	14.2	2.2	68.1
Balanus balanoides	Cyprid	37.7	14.0	3.2	51.2
Callinectes sapidus	Megalopa	252.0	6.5	2.0	37.5
Annelida					
Polydora websteri	Nectochaete	1.8	10.0	2.5	17.0

Note: Constituents are expressed as a percentage of total dry weight.

III. ENERGETICS

A. Energy Substrates During Larval Development

Studies carried out on the biochemical composition of eggs of a number of marine invertebrates suggest that protein forms the main constituent of the eggs, followed by lipid and carbohydrate.[76,78,79-82]

The high levels of protein and lipid, as compared to carbohydrate, suggest that carbohydrate is not a major energy reserve in eggs. As a result, most discussions of the energy reserves necessary for egg development have focused on the use of either protein or lipid. Needham suggested that protein was the major reserve of all aquatic eggs both freshwater and marine.[83] However, Holland noted that the eggs of marine species were poorly represented in Needham's data.[18] The use of protein reserves by two species of barnacles, *Balanus balanoides* and *B. balanus,*[84] as well as the use of protein by *Macrobrachium idella* confirm that some marine invertebrates use proteins as the major energy source for early development;[85] however, most marine eggs use lipid as the principal energy reserve.[86] Both neutral and phospholipid fractions are utilized during development of sea urchin eggs.[80,81] Dawson and Barnes suggested that the breakdown of phospholipids made a significant contribution to the energy metabolism of developing barnacle eggs.[87] Bayne et al.[88] reported that phospholipids as well as triacylglycerols were used during the development of the mussel *Mytilus edulis* from the unfertilized egg to the 7-day-old veliger larva.

Compared to the biochemical data on developing eggs of marine invertebrates, data on larvae are scarce (Table 1). Millar and Scott showed that in *Ostrea edulis* larval energy was supplied by protein, lipid, and carbohydrate reserves.[89] They concluded, however, that more energy came from lipid than from protein and carbohydrate taken together. Holland and Spencer found that the developing larvae of *O. edulis* accumulated neutral lipid that increased from 8.8% of the total organic matter in newly released larvae to 23.2% just before metamorphosis.[90] The initial growth rate of newly released *O. edulis* larvae can be positively correlated with the lipid content of the larvae

Table 2
UTILIZATION OF ORGANIC MATERIAL DURING
EMBRYOGENESIS AND NAUPLIAR LECITHOTROPHY
IN *BALANUS BALANOIDES*

Stage of development	Egg/Embryo	Nauplius I prehatching	Nauplius I swimming	Nauplius II
Total dry wt. (µg larva^{-1})	1.34[84]	0.95[84]	1.02[96] 0.80[78]	0.78[96]
Carbohydrate (ng larva^{-1})	85[84]	22[84]	8.1[96]	6.3[96]
Protein (ng larva^{-1})	750[84]	589[84]	450[96]	210[96]
Lipid (ng larva^{-1})	340[84]	301[84]	72[96]	35[96]
Respiration (nℓ larva^{-1}h^{-1})	0.2[84]	0.9[84] 0.5	7.4[96] 3.9[99]	— —
Respiration (nℓ µg^{-1}h^{-1})	0.155[84]	0.95[84]	4.9[78]	—

Note: Compiled from Holland,[78] Barnes,[84] Achituv et al.,[96] and Davenport.[99]

upon liberation.[91] Embryonic development of *Mytilus edulis* also takes place largely at the expense of the lipid reserves. The lipids acquired during development are necessary for successful metamorphosis.[88] Lipid oxidation also accounts for the significant increase in total energy metabolism during embryonic and early larval development in sea urchins, while the rate of carbohydrate metabolism does not change.[92]

The biochemical changes during embryonic development of the barnacle *Balanus balanoides* (egg through prehatching nauplius I) are shown in Table 2. These changes result in a net decrease in dry weight (29%) during embryogenesis. During the embryonic period the rate of energy metabolism increases more than four-fold. This results in an increase in the weight-specific rate of oxygen consumption and indicates a loss of storage material. All three classes of organic compounds (lipid, protein, and carbohydrate) decrease during this period.[84,93-95] Approximately 75% of the initial carbohydrate is oxidized. The loss of such a large percentage of carbohydrate suggests that carbohydrate functions as an energy reserve; however, it comprises only 5 to 15% of the total organic material in the egg. Thus, relatively minor losses of protein (25%) and lipid (10%) account for an energy input comparable to carbohydrate (25.6 nℓ O_2 egg^{-1} for protein; 23.9 nℓ O_2 egg^{-1} for lipid).[84] Importantly, the total lipid fraction decreases only 10%, but the triacylglycerols decrease 85%.[87] It is interesting to note that in this study the energy equivalent of the lost organic material does not agree with measured oxygen consumption. This probably reflects an error in the measurement of oxygen consumption. Barnes has suggested that oxidation of an organic material that was not measured might account for this discrepancy;[82] however, this seems unlikely since the organic materials that were measured accounted for 90 to 95% of the dry weight.

Following embryogenesis, the barnacle nauplius I larva is released into the plankton. It quickly molts to the nauplius II. The duration of the nauplius I — nauplius II intermolt is thought to be constant.[84] Changes that occur between the release of the free-swimming nauplius I and the nauplius II indicate the net utilization by the nonfeeding nauplius I stage. Dry weight decreases by 23% but size increases.[96] The loss of organic

Table 3
GROWTH AND METABOLISM OF *BALANUS*
BALANOIDES DURING NAUPLIAR PLANKTOTROPHY

Stage of development	Nauplius II onset of planktotrophy	Nauplius VI completion of planktotrophy
Total dry wt. (μg larva^{-1})	0.78[96]	14.2[78]
Respiration (nℓ larva^{-1}h^{-1})[a]	3.8[a]	32[99]
Respiration (nℓ μg^{-1}h^{-1})[b]	4.9[78]	2.3[78]

[a] Calculated using N I weight-specific rate from Holland.[78]
[b] Assumed N I = N II.

Compiled from Holland,[78] Achituv et al.,[96] and Davenport.[99]

material and increased respiratory rate corresponds to the increase in activity associated with swimming. The nauplius I utilizes protein (53% of initial) and lipid (51% of initial) extensively, while it uses carbohydrate (22% of initial) much less. Energetic potential in the organic material could sustain the larva for about two days, at measured rates of metabolism. Thus energy reserves could fuel the nauplius I — II transition and considerable reserves may be carried over to the nauplius II. The absence of food early in the second naupliar stage may not be fatal (Table 2).[96]

Energetic balances (feeding rates, metabolic expenditures, etc.) during the planktotrophic phase of barnacle development have not been published. However, there must be a net gain of energy from larval feeding, because nauplii grow (increase in size, dry weight, and organic constituents, Table 3) and are capable of storing large amounts of energy for the later nonfeeding cyprid stage. Dietary lipids are taken up by the anterior midgut cells in later larval stages (i.e., nauplius VI) and are concentrated as cytoplasmic droplets in both the midgut cells and in special oil cells surrounding the midgut.[97] The molt to the cyprid stage and thus the termination of naupliar planktotrophy seems to be sensitive to the level and rate of this lipid accumulation.[98] Subthreshold lipid levels increase the duration of the molt from the D_2 phase of ecdysis. Depletion of the oil cells in nauplius VI prior to D_2 (through starvation) prevents the sixth naupliar molt.

Development from the nauplius II to the nauplius VI involves considerable increases in dry weight (0.78 to 14.2 μg larva^{-1}).[78,96] The later larval stage has a higher energy demand.[96,99] However, the weight specific rate decreases, suggesting the net accumulation of metabolically inactive energy stores (lipid) (Table 3).

Although a variety of energy substrates are utilized by larvae, lipid seems to predominate in many types of larvae.[78] Inherent advantages exist in storing lipid. It has an oxygen demand of approximately 2×10^3 mℓ O_2 g^{-1} compared with 1.2×10^3 mℓ O_2 g^{-1} for protein and 0.8×10^3 mℓ O_2 g^{-1} for carbohydrates, and it provides nearly twice as much energy per unit weight.[100] Lipids may also provide molecular water, as well as confer to the larva some degree of buoyancy, but as Sargent has pointed out it is unclear if these processes are important to the larva.[101]

B. Utilization of Biochemical Components During Metamorphosis
Termination of the larval phase involves radical morphological and physiological changes during the transition to the adult form. These events are collectively termed metamorphosis.

Holland and Spencer suggested that neutral lipid is used as the main energy reserve during metamorphosis of the larvae of *Ostrea edulis*.[90] This is in contrast to the adult in which polysaccharide served as the major energy reserve. Using their data and standard conversions for complete oxidation of lipids, it is calculated that the cost of metamorphosis is 1.7×10^{-3} calories per larva.

Neutral lipid is the main energy source during metamorphosis of the veliger larvae in 4 species of snails, *Littorina littorea*, *L. saxatilis*, *L. obtusata*, and *L. neritoides* and in nonfeeding cypris larva of *Balanus balanoides*.[102] Waldock and Holland demonstrated that the mean level of triacylglycerols was 8% of the ash-free dry weight and fell to 2% of the ash-free dry weight during metamorphosis in the newly settled nonfeeding juveniles.[103] Lucas et al.[64] showed that during metamorphosis of cypris larvae of *B. balanoides*, lipid supplied most of the energy although some protein was used. The cost of metamorphosis was calculated at 3.0×10^{-2} calories.[64]

The nonfeeding barnacle cyprid is initially rich in organic materials. Biochemical composition shows that lipid is the major organic component. The lipid is primarily neutral lipid,[63] mostly triacylglycerols.[103] Sterols are present in small amounts; free fatty acids, diacylglycerols, and wax esters are absent. This is in contrast to stage I and II larvae in which there are considerable amounts of free fatty acids.[96]

The cyprid has significant amounts of polysaccharides, free sugars, and proteins. Holland and Walker showed that the swimming activity of the larva utilizes only the neutral lipid fraction of the organic material.[63] The formation of lipid reserves presumably occurs from dietary precursors during planktotrophic naupliar development. The biochemical composition of phytoplankton shows mostly protein and carbohydrate, with some lipid. Most larvae can presumably synthesize all saturated and monounsaturated fatty acids from carbohydrates. Only the polyunsaturated fatty acids are a dietary requirement.[78]

The weight specific metabolic rate for planktonic cyprid larvae is low, while its stored energy density is high.[64,78] This indicates a large amount of metabolically inactive material, which represents a savings of organismal rates of metabolism. Thus, while the cyprid generally settles in a few days, it has the potential to survive in the plankton for many weeks.[64]

Settlement inducing factors utilized by cyprids are specific, and it is possible to prolong the swimming phase in the laboratory. During the delay, there is a reduction in neutral lipids. Other biochemical components decrease slightly or remain constant.[63]

At the time of settlement, larvae explore the substratum for a short period of time. The energy demand during exploration is less than during swimming. Subsequent to settlement, molting occurs followed by calcification of the juvenile shell. After several days, feeding begins. Rates of oxygen consumption increase dramatically during molting and calcification suggesting that there is a substantial energy cost to metamorphosis. Lucas et al.[64] measured rates of oxygen consumption and calculated a total demand of 6.7 $\mu\ell$ larva^{-1}. Energy is supplied by the oxidation of the triacylglycerols, which are reduced from 8% to 2% of the dry weight during metamorphosis. All classes of triacylglycerol fatty acids are utilized non-selectively.[103]

Analysis of biochemical components and measurements of oxygen consumption rates of cyprid larvae of *Balanus balanoides* maintained in the laboratory showed that although lipid is the primary energy reserve, protein is also utilized.[63,64,103] During metamorphosis, neutral lipid decreases to 9.6% of the total organic matter and then remains constant up to 25 days after settlement.

Day found that larvae of the spionid polychaete *Polydora websteri* accumulates lipids during development.[76] During metamorphosis there is a significant decrease in the amount of lipid present. The greatest loss of lipid is due to the neutral lipid fraction.

The caloric cost of metamorphosis of *P. websteri*, assuming total oxidation of the lipid fraction, is 1.2×10^{-3}. The caloric cost of metamorphosis of the cyprid larvae of *Balanus balanoides* is an order of magnitude higher than either *P. websteri* or *Ostrea edulis*. However, Lucas et al.[64] actually calculated the energy requirements for settlement, one molting cycle, and initial shell calcification. It is not therefore surprising to find a larger energetic cost.

A somewhat different pattern holds for the opistobranch mollusk, *Aplysia juliana*. The veliger stores lipid during larval development;[104] however, proteins rather than lipids are utilized during metamorphosis. This is in keeping with the resorption or ingestion of larval tissues that are lost during molluscan metamorphosis.[105,106] Kempf (personal communication) hypothesizes that lipid reserves might be utilized by the early post-metamorphic juveniles, if food was scarce in the habitat in which they settled.

C. Delay of Metamorphosis

It is generally understood that the end of larval life includes two steps: settlement, which is a reversible behavior; and metamorphosis, which is not reversible and results in a change of body form to an adult.[4] In the absence of a suitable substratum, most larvae are capable of prolonging their larval life, thus postponing metamorphosis.[22,66,107-112] Generally, the delay of metamorphosis results from lack of a suitable substratum responsible for the induction of metamorphosis. The substratum may be a particular kind of alga, sediment with a certain texture, or a particular food item.[4,62]

Specific induction of metamorphosis may provide an increased probability of the larva encountering a suitable substratum for metamorphosis. It has been suggested that duration of the delay phase is in large measure determined by the ability of the larva to enter into an energetic steady state, in which net energy intake is equal to basic metabolic demands. When a balance exists, there is no excess energy available for growth or further differentiation of larval tissue.[113-115] This steady state allows the larva to maintain a specific level of organization for a specific length of time.[66] The evidence for such a state is primarily indirect.[76,114,116] Recent work by Kempf has provided evidence for a steady state in terms of mass during extended larval life in competent veligers of *Aplysia juliana*.[117] However, there is evidence tht some species continue to grow and differentiate during the competent period, suggesting that a metabolic steady state is not universal.[116,118]

Relative delay capability and the fate of larvae that do not encounter a suitable substratum varies among species of marine invertebrates. Competent larvae of the polychaete, *Spirorbis borealis*, are not able to feed in the plankton and can delay metamorphosis for only about 12 hours before losing the ability to metamorphose successfully.[119,120]

The ability to delay settlement is adaptive only if it eventually results in successful metamorphosis. Lucas et al.[64] found that barnacle cyprids that delayed for more than a few weeks were not capable of successful metamorphosis.

Bayne showed that the potential length of the delay phase of *Mytilus edulis* larvae would vary with temperature, salinity, and pH of the culture medium.[66] It has been hypothesized that the length of the delay period may be controlled by energy balance considerations, in at least some species, although larval energy budgets have not been determined during the delay of metamorphosis of any marine invertebrate.

D. Starvation

Although effects of starvation on survival of planktotrophic larvae are not well known, a few studies indicate that the larvae are able to withstand long periods of starvation without increased mortality.[66,90,121,122] Thorson and Vance have hypothesized that reduced food levels could lengthen developmental time, increasing the susceptability of larvae to planktonic predation.[3,19] If developmental rates are influenced by environmental conditions such as temperature or food availability, these factors should correlate with the length of larval life in the plankton.

Holland and Spencer demonstrated that phospholipid is lost during starvation of *Ostrea edulis*.[90] The loss of phospholipid was interesting because phospholipid made an important contribution toward energy metabolism during the development of the larva. Holland and Spencer made no mention of the loss of phospholipid as it related to successful metamorphosis.[90]

Day found that duration of the starvation period was not the only important factor determining the effects of starvation on larvae of the polychaete, *Polydora giardi*.[123] Timing of the starvation period also played an important role. He suggested that there was a "critical period" in the larval development of *P. giardi* during which the larvae were more susceptible to low food levels than at other periods during development. This has also been demonstrated with crab larvae.[73]

These conclusions are similar to those reached by Shelbourne,[124] Lasker,[125] Wibory,[126] and others, who suggest that a critical feeding period exists in some fish larvae. If fish larvae do not feed immediately following the absorption of their yolk sac, they die. The critical period found in *P. giardi* larvae appears to be at release from the egg capsule. At this point in development they have utilized most of the stored reserves supplied by the adults, and must feed to survive and develop. Larvae of *P. giardi* appear to store energy during development, and if they encounter low food levels after the critical period, they are able to utilize the stored energy. Stored energy in the form of lipids is lower in larvae which are forced to develop under conditions of low food availability. It was hypothesized that lower levels of lipid would significantly affect survival rates of the larvae during and after metamorphosis.[76]

Starvation of parents can affect larvae. Helm et al.[91] reared the larvae of *Ostrea edulis* from parents that were held under two different feeding conditions in the laboratory. They measured the "vigor" of the larvae, larval growth rates, and survival of metamorphosed juveniles. Larvae from adults which were held in unsupplemented seawater had slower growth rates than larvae from adults reared in seawater supplemented with phytoplankton. The slower growth rates were significant because the rate of larval growth over the first 96 hours was itself predictive of spat yield. Larval vigor was correlated significantly with the total neutral lipid content. This indicates that animals under nutritive stress can and do produce gametes, but these gametes may develop into embryos and larvae that are less viable than embryos produced by adults not under nutritive stress.

IV. DISPERSAL

A. Duration of Pelagic Period and Scale of Dispersal

The scale of dispersal depends on movement of the water masses, larval behavior, and duration of the pelagic stages.[127-130] Larval longevity and water movements establish the potential for dispersal, whereas larval behavior often determines the actual degree of spread. Two independent means have been employed to estimate the quantitative relationship between duration of the pelagic period and scale of dispersal.[130] Extensions of geographic ranges in species with pelagic larvae and sedentary or sessile adults suggest that a 10 to 15 day larval period results in spread over 20 to 30 km.[131-133] These estimates are in agreement with oceanic diffusion measurements.[134] Diffusion data suggest a positive correlation between time and spread (Table 4).[135] Several hours result in spread on the order of hundreds of meters, 1 to 2 days allow dispersal over

Table 4
OCEANIC DIFFUSION[130]

Pelagic life	Log$_{10}$ (probable distance transported in cms)	Order of magnitude	Whether likely to be exceeded by tidal currents
3 to 6h	4	100 m	Yes
1 to 2 days	5	1 Km	Yes
7 to 14 days	6	10 Km	?
14 days to 3 months	7	100 Km	No
1 year	8	1000 Km	No

approximately a kilometer, a week or two increases the scale to 10 km, 2 weeks to 3 months correspond to 100 km, and a year can result in movement of 1000 km.[134] These estimates are valid for surface waters without appreciable tidal or residual currents. Crisp has argued that tidal currents exceed the scale of diffusion in some coastal regions over short periods of time (hours to several days).[5,131] Tidal flows result in a cyclic pattern of translation with the first maximum at approximately 6 hours. Successive tidal exchanges do not appreciably increase the scale of spread. Significantly greater dispersal requires a substantially longer pelagic period, probably on the order of one week.[5,135]

It was previously believed that the potential for transoceanic dispersal was very rare.[129,136,137] Thorson argued that 80% of the planktonic larvae lived less than 6 weeks in the plankton.[129] Given maximum estimates for major oceanic currents, there was simply no chance for transoceanic larval dispersal. There were only a few cases of exceptionally long-lived larvae with long distance dispersal potential (e.g., some crustaceans and echinoderms).

Scheltema provided evidence that pelagic larvae of most major invertebrate taxa were continuously distributed across the Atlantic Ocean in the major currents that comprise the North Atlantic Gyre and in the major undercurrents.[113,114] He described these larvae as teleplanic, defined as (1) originating from shoal-water, continental shelf benthos; (2) found regularly in the open sea; (3) with larval stages of long duration; and (4) serving as a means of long distance dispersal. Scheltema argued that Thorson's estimates were biased towards cold temperate species with rapid development and short pelagic periods.[114]

The ability to delay settlement and metamorphosis and to persist as a planktonic larva for extended periods is suggested as an important means of increasing the likelihood of successful transport across ocean basins. The delay phase, at least in the gastropod *Cymatium*, was thought to involve a metabolic steady state without growth.[113] Recently, Pechenik and Kempf have provided evidence that gastropod larvae may live in the plankton for a long time after achieving a metamorphically competent state.[117,118]

B. Types of Spread

Planktonic larval stages are typically the major dispersive agents in mixed (benthic/planktonic) life cycles. The dispersal of larvae is therefore an important determinant in the recruitment and distributional patterns of benthic adult populations.

There are two dissociable components to dispersal of meroplanktonic larvae of marine invertebrates. First is the spread from the benthic parental population. Second is the spread amongst sibling larvae (i.e., offspring from the same parent).[135]

Spread from the parental population is the central concern in analysis of range extension, transoceanic dispersal and retention in estuaries.[4,5,113,130,131,138-140] Dispersal from siblings is potentially associated with averaging the risk to a given parent in temporally/spatially variable environments.[141-145] Dispersal from a point source (parent) depends on such factors as time in the plankton, current direction, and current velocity. Spread of siblings is related to variation in spawning time (release of larvae into the plankton), vertical distributions of larvae in the water column, variation in length of the larval phase, and the total length of the larval phase.[8,135]

The potential mechanisms for increasing spread of sibling larvae are reviewed in detail by Strathmann.[135] The present treatment will be restricted to a few brief comments. The spread of siblings is increased with longer pelagic periods, but the relative spread effected by each incremental increase in planktonic duration decreases.[135] Prolonging the pelagic period is most effective in species with a relatively short larval phase.

Differential vertical distribution should increase horizontal spread, as should variation in the duration of the larval stages. The effectiveness of these mechanisms should depend on the degree of variation relative to the total larval period. Horizontal spread should be greater for any given degree of vertical variation in proportion to the duration of the pelagic period.

Multiple spawnings should become less effective for spreading sibling offspring as the planktonic periods become longer.[135] It must be stressed that although examples of these patterns are numerous, there is little evidence to support a casual connection between natural selection for these traits and the advantages conferred by the spread of sibling larvae. Furthermore, the above discussion applies only to instances where spread of siblings is to be maximized; this is not a universal feature of marine invertebrate life cycles.

C. Larval Behavior and Retention of Larvae in Estuaries

The actual distance that larvae are dispersed in the plankton may be influenced by swimming behavior. Active control of vertical position in the water column can be an effective means of minimizing or maximizing the degree of horizontal transport within the limits established by water movement and the duration of the planktonic period.[146]

Recruitment considerations set the appropriate scale of dispersal. Recruitment into a benthic estuarine population requires that larvae or juveniles remain within the parental population or return to it.[147] Species with planktonic larvae must rely on either larval transport and settlement in the region of the benthic population or migration of juveniles from marine populations.[138]

Examples of larval behavior controlling dispersal are found in many estuarine species with a long planktonic phase.[138,140,146-149] The net nontidal flow in an estuary is seaward. This results in directional transport and exposure to a gradient of salinity. The seaward surface flow is countered by a deeper landward drift. Bousfield provided evidence that the vertical distribution of larvae could control the horizontal transport of larvae within estuaries.[140]

Wood and Hargis claimed that vertical distribution during a tidal cycle was actively controlled by the larvae based on distributional differences between passively transported particles and oyster larvae.[148] This idea has been challenged.[139,150] Based on field and laboratory studies of barnacle cyprids, de Wolf has claimed that passive transport by currents can account for the distribution of the larvae.[150] It was also claimed that the differences between passive particles and larvae observed by Wood and Hargis could be explained by differences in size and density and that sufficient evidence did not exist to reject the hypothesis that vertical distributions of larvae are determined primarily by water movements.[139] Presently, most larval biologists accept the idea that larvae (at least strong swimmers such as crustacean larvae) can control their vertical position in the water column and thus influence horizontal transport.

Laboratory studies have demonstrated that larvae respond to various environmental factors such as light,[59,151-153] gravity, and hydrostatic pressure.[154-157] Responses include directional swimming (e.g., phototaxis and geotaxis) and changes in swimming velocity (e.g., barokinesis).

The behavioral responses of larvae to environmental factors are determined by the particular requirements for larval survival and recruitment of each species. The xanthid mud crab, *Rhithropanopeus harrisii*, is a common estuarine species along the central and southern Atlantic coast of the United States. It represents one extreme in the range of recruitment patterns found among benthic estuarine crabs. The adults are restricted to estuaries (0.006 to 25 ppt salinity), and the most important means of recruitment is retention of larvae within the estuary.[44,146] Field studies suggest that larvae control their vertical distribution and maintain their position by moving up the estuary with deep currents.[146,147]

Rhithropanopeus harrisii develops from hatching through metamorphosis to the juvenile crab in 18 days (25°C, 25 ppt).[158] Larval development consists of four zoeal stages and one megalopa stage. In contrast to most estuarine crabs, hatching of zoeae occurs in low salinity water.

Although the juveniles and adults are restricted to low salinity waters, experiments indicate that the larvae of *R. harrisii* tolerate a wide range of environmental factors.[44] Thus, it would be possible for larvae to survive transport out of an estuary into marine conditions. Dispersal of this type is enhanced in many crustaceans by variation in the number of larval instars amongst siblings.[49] *Rhithropanopeus harrisii* however seems to have a strictly determined number of larval stages.[44] This may serve to minimize horizontal spread from the parental population.

The mechanisms of depth regulation in the larval stages of *R. harrisii* involve complex behavioral responses to several physical environmental factors, superimposed upon an endogenous tidal rhythm of vertical migration.[159] The zoeal stages respond to light with positive and negative phototaxes,[58,160,161] as well as polarized light.[162] The phototactic responses are modified by temperature,[163] salinity,[154] food concentration, and stage.[160,164]

In addition to the phototactic behavior, larvae also respond to gravity (geotaxis)[163] and also to changes in hydrostatic pressure (barokinesis) on the scale that they are likely to encounter in the field.[156] Barokinesis provides a fine tuning of the directed responses to light and gravity.[155,156] The means by which larvae integrate these internal and external factors to effect a tidal rhythm of migration is not known.

It seems that these behaviors of *R. harrisii* are not designed to reduce planktonic mortality by avoidance of physical-physiological stress. Larvae can tolerate fully marine conditions. Rather the adaptive value lies in enhancing recruitment through minimizing dispersal.

In addition to controlling vertical position in the water column, the zoeal stages have a phototactic predator avoidance reaction. This behavior does seem to have immediate survival value for the planktonic stages.

It is clear that the planktonic stages of the life cycle of *R. harrisii* possess adaptations of both immediate significance (i.e., avoiding predation in the plankton) and others that are designed to solve problems of recruitment to the benthos. It is not surprising that even complex life cycles are functionally integrated sets of phenotypic traits. It will be necessary to interpret the behavior of each meroplanktonic species in light of the ecology of the benthic stages. Although holoplanktonic species utilize a diverse array of behaviors to maintain position in estuaries the differences can be much more pronounced in meroplankton because of the potential for benthic migration by juveniles and adults.[165] This is seen clearly in the comparison of *R. harrisii* which must retain its larvae within the estuary for recruitment and *Callinectes sapidus* which utilizes offshore recruitment. *Callinectes sapidus* responds to the same environmental factors as *R. harrisii* but in very different ways. The result is a mechanism for exchange of larvae between the estuary and the coastal environment.[149,166] Thus, two species that inhabit the same environment and have similar larval forms behave quite differently. This is not because of differences in the tolerances or requirements of the planktonic stages, but rather because of different ecological problems faced by the benthic adults.

D. Precompetent and Competent Periods

Meroplanktonic larvae eventually settle to the bottom, metamorphose into the adult form and assume a benthic, often sessile, existence. Many aspects of the planktonic phase of the life cycle must be interpreted in light of the planktonic-benthic transition.[13]

Settlement behavior is a complex ecological phenomenon. The interested reader is referred to major reviews.[4,62,107,110,167,168] For the present discussion it is sufficient to emphasize two points. First, settlement can be highly specific, the larvae exerting a choice with regard to the substratum upon which they eventually settle. Cues to settlement include chemicals, surface texture, illumination, presence or absence of conspecific adults and juvenile or adult predators or prey. Second, settlement is a reversible behavior. The transition to the benthos can be delayed, and the planktonic phase can be extended appreciably.

A larva that is developmentally capable of settlement and metamorphosis is called metamorphically competent. Competency is a physiological state that is independent of substratum availability or behavioral cues. It depends on morphogenesis of presumptive adult/juvenile structures and nutritional state.

The planktonic larval phase can be divided into a precompetent and a competent period. Duration of the precompetent period depends on rates of growth and development. It is thus modified by such factors as genetic variation in growth rate, temperature, and planktonic food availability. It is not clear to what extent present estimates for minimum duration of the precompetent period reflect inherent variation or suboptimal culture conditions.

A competent larva usually requires a stimulus to metamorphose. Lack of an appropriate stimulus often leads to a period of delay.[62,110] The delay period can be quite long, both relative to the precompetent period and in real time. The ascidian, *Chelyosoma productum*, is capable of settling in approximately 2 hours following release from the parent, yet it can delay for up to 48 hours.[169] The opistobranch mollusk *Aplysia juliana*, is competent after approximately 30 days. In the absence of a metamorphic inducer, larvae continue to swim and remain competent up to 311 days.[117]

Factors controlling delay capability are not well-known. Energy can be an important limiting factor in nonfeeding larvae (e.g., barnacle cyprids).[4] Lucas et al.,[64] and Holland and Walker[63] showed that cyprids can survive on stored materials during a non-feeding delay period up to 6 to 8 weeks. However, success at metamorphosis and survival of the juvenile until the onset of feeding decreases drastically at the end of three weeks of delay. Lucas et al. demonstrated that metamorphic events are energetically expensive and severely limit the effective delay period.[64]

It is not, however, a safe generalization that all nonfeeding larvae are restricted to a short competent period by limited energy reserves. Birkeland et al. found that the lecithotrophic larva of the starfish, *Mediaster aequalis*, reach competency in approximately 30 days but can successfully metamorphose after 12 months of delay.[170] Hadfield (personal communication) has found that the opistobranch mollusk, *Phestilla sibogae*, which has a normally nonfeeding larva, can facultatively feed in the plankton during metamorphic delay.[171] *Phestilla sibogae* larvae are competent to metamorphose

4 days after hatching. The delay period can be up to 6 weeks by means of larval feeding.

Obrebski, in a model of larval colonizing strategies, suggested that the delay of metamorphosis has associated costs and benefits.[172] The benefits derive from the increased likelihood of encountering suitable substrata. The primary cost is in allocation of energy to habitat search, thereby limiting the energy available for metamorphosis. The relationship between delay of metamorphosis and postmetamorphic survival or juvenile growth, although intuitively reasonable, has never been demonstrated.

The delay capability of feeding larvae should not be energy limited. This does not imply that there is an infinite delay potential. Many organisms have only a limited delay phase, often with substratum specificity decreasing with time.[4,119] The factors controlling delay capability are not well-known. Senescence of larval tissues may be important in some larvae, but more information is needed.[66,116,117,173-175]

Pechenik has hypothesized that there may be a program of larval development with a genetically fixed endpoint.[118] He suggested that the rate at which a larva completes the precompetent period directly determines delay capability; slowly developing larvae have long competent periods. Selection for long competent periods must therefore decrease larval growth rates. Functionally (ontogenetically), the precompetent period determines the competent period; however, this implies that in evolutionary terms the demand for a given competent period would determine the precompetent period. The gastropod larvae that Pechenik studied continued to grow during the competent period.[82,118] Other studies on molluscan larvae suggest that the delay period involves a metabolic steady state without growth nor differentiation.[105,113,117,176] Clearly, a better understanding of the physiology of larval growth and differentiation is required to elucidate the control of competency and metamorphic delay capability.

Jackson and Strathmann have provided an alternative explanation for the relationship between the length of the precompetent and the competent periods.[177] They proposed that a long competent period was necessary to balance a long precompetent one. Their eddy diffusion model showed that offshore-onshore mixing may transport larvae away from suitable settlement sites during the precompetent period, when settlement is not possible. To balance offshore transport (i.e., to increase the likelihood of transport back to suitable settlement sites) larvae should have a long delay period. From this perspective, natural selection should act on the competent phase to offset precompetent transport.

E. Consequences of Dispersal

The adaptive significance of pelagic dispersal is assumed to be the result of: (1) an increase in the geographical range of the species; (2) the ability of the species to colonize new areas or occupy ephemeral habitats; and/or (3) increased genetic exchange over a large population.[114,130,178,179]

F. Geographic Range

Thorson suggested that organisms which possessed highly dispersing larvae were successful over large geographical ranges.[22] Scheltema has argued that successful "drift and settle" sequences could lead to trans-Atlantic journeys in several invertebrate forms with long pelagic larval periods.[113,114,180,181] He predicted that as a result of their dispersal capacity they should be broadly adapted species with considerable genetic diversity.[128] If some populations within the species range became extinct, such areas could be repopulated by larvae from distant regions when suitable conditions are restored. Recently, the significance of dispersal in establishing and maintaining biogeographic ranges has been questioned.[182]

G. Colonization

Scheltema has suggested that dispersal allows organisms to find and colonize new habitats.[179,181] It would seem logical to suppose that natural selection would clearly favor individuals which leave crowded habitats and successfully colonize empty habitats and subsequently leave many descendants. Indeed, the concept of achieving some advantage as a result to the ability to disperse, or compensating in some manner for the lack of dispersal forms the foundation of many life history theories.[183]

The deterministic model of r and K selection assumes that organisms have two strategies available to them as a result of their ability to disperse and to compete once they become established. It is predicted that species with a high degree of dispersal are able to colonize new habitats while those species with poor dispersal have evolved a competitive advantage allowing them to persist in the environment.[184] Life history models based on stochastic events predict that, in fluctuating environments, species with high juvenile mortality will evolve patterns of reproduction which limit dispersal. Environments in which adult mortality is high will be dominated by species which have the ability to disperse.[183]

Pelagic dispersal is associated with high fecundity to offset mortality experienced during the planktonic period.[130] Williams' "Elm-Oyster Model" emphasizes an analogy between pelagic larvae and wind-borne seeds.[185] He argued that intense competition in a habitat would result in high fecundity and wide dissemination. Spread of larvae would increase the opportunity to colonize habitats which differ from those of the parent. Crisp has argued that larval transfer is an essential process for the survival of species which occupy isolated transient habitats subject to density independent extinctions.[5,62,130]

For a dispersing larva to be successful, acceptable environmental conditions for the larva must be available. For example, there must be areas available for colonization within the range of spread from the adult habitat. A great deal of effort has been expended in determining the life span of the larva as well as the direction and possible distance of dispersal.[114,117,135,181,186] These studies have demonstrated that some species of marine larvae live long enough to be transported by ocean currents to other continents or remote islands. It has been considered that given enough opportunities and a large number of larvae, long distance colonization is not only possible but probable. For example, Dana has suggested that while most genera of hermatypic corals disappeared from the Eastern Pacific during the Tertiary, this region has been colonized by transoceanic planula larvae from the Indo-West Pacific.[187]

Objections have been raised, questioning the validity of hypotheses which suggest that long distance dispersal aids colonization. Strathmann has stated that many organisms with pelagic larvae have life cycles which do not suggest a colonizing strategy.[135] In many species of benthic invertebrates large scale dispersal always accompanies reproduction. Larvae settling in a favorable area subsequently disperse their offspring such that only a few of the offspring are recruited to the parental area. As a result, these species cannot take advantage of favorable local conditions to increase population size over several generations. Long-term dispersal would be advantageous if conditions were always deteriorating near the parental population.[188] Under such conditions the probability of offspring survival is always greater at a distance from the majority of spawning adults. However, many benthic invertebrates do not show extreme patchiness, and it is unlikely that probability of survival is always higher away from spawning adults.

Large-scale dispersal often does not result in escape from crowded conditions, in fact settling behavior often results in dense aggregations. Larvae often are brought together by specific substratum preferences.[107,189-191] Thus, spacing of individuals or avoidance of crowding does not appear to be a function of a long distance pelagic dispersal.

An increased probability of encountering a favorable settling site is not necessarily a result of a long dispersive stage. Many larvae develop in the plankton for many weeks before they are capable of settling and cannot take advantage of good sites during this period. A larva which could travel a great distance over the bottom in the latter part of the pelagic stage, during which time it could settle, would be more likely to encounter favorable settling sites.

One would expect that increased substratum selection would be one of the advantages of dispersal, but if it were the only advantage of a pelagic stage, one would expect that larvae would not be released until they were able to settle. Fauchald has noted that polychaetes which are often the most successful in terms of colonization, often lack a dispersal stage.[192]

There are examples of species widely distributed in both time and space which lack pelagic stages, such as the periwinkle *Littorina rudis* and many interstitial animals. This implies that the relationship between dispersal and colonization is not entirely clear.

H. Gene Flow

In species with limited dispersal, greater variation should exist among populations. In the long run this should favor local speciation. Conversely, species which disperse over wide regions should retain specific status over large areas.

The ecological consequences of gene flow between local populations by pelagic larvae of marine benthic invertebrates have been modelled by Crisp and Scheltema.[5,127,128,130] It is generally agreed that dispersal capability establishes the limits of the panmictic population and therefore prevents adaptation to local conditions.[193] This consequence of larval dispersal can be detrimental, as in increased mortality due to a lack of local adaptation.[130] For example, barnacle larvae disperse over a greater area than that in which settlement cues are reliable. Inappropriate settlement results in considerable mortality.[194]

Scheltema predicted that marine species with long distance larvae would be broadly adapted species with large genotypic variability, broad geographic range and great temporal longevity.[128] Species with less dispersal capability should show geographic variation and varying degrees of speciation. He suggested that species without a dispersal stage must either change with the environment or become extinct.

The ability to distinguish planktonic and nonplanktonic modes of larval development in gastropod fossils has led to analysis of the relationship between dispersal capability and rates of evolution.[127,128,195-199] Species with planktonic larvae seem to persist longer in geologic time than species without planktonic larvae (4.4 million years versus 2.2 million years for Paleocene-Eocene volutid gastropods).[198,199] The dispersal capability conferred by pelagic development also correlates with wide geographic ranges within any short period of geologic time.[127,198,199] Extremes in the range of species longevity result from wide environmental tolerances (eurytopy) with planktonic dispersal or narrow tolerance (stenotopy) with limited dispersal.[127,198,199] Dispersal is thus viewed as a means of maintaining gene flow, and of preventing isolation and divergence of populations, and thereby reducing speciation rates. Eurytopy minimizes the likelihood of extinction. The combination yields a widely dispersed, slowly evolving species.

Levins has argued that dispersal counteracts the effects of large scale unpredictable events by maintaining the geographical range of a species wherever and whenever favorable conditions exist, thereby avoiding extinction of species.[200] Dispersal, by providing genetic continuity among separated populations, should maintain a high genetic variability within the species. Levins has concluded that the adaptive significance of dispersal is that it permits populations to respond to long term widespread fluctuations in the environment while reducing the effects of small scale environmental changes. Dispersal also tends to decrease geographic variability.[200] Lewontin suggested that a migration rate as small as one individual in a thousand per generation is sufficient to prevent differentiation.[201] MacArthur and Wilson suggested that species confined to environments which lack temporal heterogeneity have reduced dispersal ability and have small population sizes.[184] These areas were considered to be species rich and made of communities with high diversity. Based on these observations, Grassle predicted low genetic variability in the deep sea.[202] This hypothesis has been shown to be false for a number of species.[203,204] Ayala et al.[205] and Valentine have explained increased variability by the trophic stability hypothesis.[206] This hypothesis appears applicable to areas of seasonal upwelling of nutrient-rich waters. Opportunistic responses to this nutrient source should lead to effective dispersal to ensure high genetic variability.

There are also serious objections to the assumption that species with greater dispersal ability tend to be less divided into genetically different subpopulations.[207] A number of exceptions to this generalization have been discussed by Koehn.[208] Of greater concern is the possibility that allelic variation of loci measured electrophoretically may bear little relationship to the variability of the genome as a whole. If indeed this is the case, it would require a reexamination of the evidence for the advantages of long-distance dispersal in relation to geographical range and genetic exchange.

The preceding discussion on the advantages of dispersal assumes that dispersal is of selective advantage based on the increased advantages of colonization, increased geographical range or increased genetic exchange within a species population. These advantages have been generally accepted as axiomatic but in most cases remain untested.

It has been suggested that large scale dispersal resulting from a long pelagic life cycle may be of no advantage, but rather may be the accidental result of a pattern of development favored for other (e.g., trophic) reasons. It has been proposed that what may be more important is short distance dispersal.[135] This implies that a small-scale dispersive ability, with offspring of varying genotypes, would be favored when the quality of areas vary over short distances.[135] It is well-known that larval forms from single pairings are extremely variable in terms of growth rates, developmental times, and degree of substratum selectivity.

By means of computer simulation models, Palmer and Strathmann examined the relative advantage of increasing scales of dispersal.[141] There are significant advantages to dispersal when compared to the lack of dispersal. But increasing the scale of spread does not, in general, yield increasing advantages. They suggested that the long pelagic periods associated with planktotrophic larvae are probably not the result of selection for large-scale dispersal. Rather the advantages of planktotrophy over lecithotrophy, with a short pelagic period, are more likely to be trophic. The differing scales of dispersal are consequential, not casual, factors in the evolution of these life history patterns.

V. CONCLUSIONS

The existence of a planktonic larval stage in the life cycle of a marine benthic invertebrate is an evolutionary option. Consequently, larval biologists are concerned with the adaptive consequences of pelagic and nonpelagic development. In general, it seems that the pelagic realm offers trophic and dispersal opportunities not available to benthic organisms.

Ways in which larvae are specialized for a planktonic existence are known in detail for very few larval types. How do planktonic larvae capture food, respond to prey availability, swim, orient, select substrata, defend against predators, grow, respond to environmental stresses, and disperse? There are no general answers yet. Indications are that larval forms are so diverse taxonomically, functionally, ecologically, and evolutionarily that general answers may not be forthcoming. Almost every major taxon contains some species with pelagic larvae. It is unlikely that there will be more than superficial generalizations that are valid for such an assemblage of organisms.

By definition, meroplanktonic organisms are temporary residents in the plankton. Meroplanktonic organisms must integrate pelagic and benthic stages of their life cycles. An individual organism (e.g., starfish), as a function of time, may be a beautifully adapted herbivorous planktonic larva and then an equally specialized benthic carnivore in the intertidal zone. To be successful at one stage of the life cycle requires a successful transition between all of the stages. Gametogenesis, spawning, and fertilization by the benthic stages are important determinants of initial planktonic success. Likewise, settlement behavior by pelagic larvae determines the habitat of the benthic adult. We believe that it is this property of life cycle integration that establishes the proper perspective for the analysis of the meroplanktonic component of the marine zooplankton community.

ACKNOWLEDGMENTS

We would like to thank the following reviewers for their comments on the manuscript; Dr. J. Simon, Dr. R. R. Strathmann, Dr. S. D. Sulkin, Dr. S. Kempf. We would also like to thank Dr. A. O. D. Willows, who provided facilities at the Friday Harbor Laboratories, University of Washington.

REFERENCES

1. Chia, F. S., Classification and adaptive significance of developmental patterns in marine invertebrates, *Thalassia Jugosl.*, 10, 121, 1974.
2. Hardy, A., *The Open Sea: Its Natural History,* Collins, London, 1958.
3. Thorson, G., Reproduction and larval development of Danish marine bottom invertebrates, *Medd. Komm. For. Dan. Fisk. Havunders.,* 4, 1, 1946.
4. Crisp, D. J., Energy relations of marine invertebrate larvae, *Thalassia Jugosl.,* 10, 103, 1974.
5. Crisp, D. J., The role of the pelagic larva, in *Perspectives in Experimental Zoology,* Spencer-Davies, T., Ed., Pergamon Press, 1976, 145.
6. Mileikovsky, S. A., Types of larval development in marine bottom invertebrates, their distribution and ecological significance: a reevaluation, *Mar. Biol.,* 10, 193, 1971.
7. Mileikovsky, S. A., Types of larval development in marine bottom invertebrates: an integrated ecological scheme, *Thalassia Jugosl.,* 10, 171, 1974.
8. Vance, R. R., On reproductive strategies in marine benthic invertebrates, *Am. Nat.,* 107, 339, 1973.
9. Vance, R. R., More on reproductive strategies in marine benthic invertebrates, *Am. Nat.,* 107, 353, 1973.
10. Underwood, A. J., On models for reproductive strategy in marine benthic invertebrates, *Am. Nat.,* 108, 874, 1974.
11. Perron, F. E., Larval biology of six species of the genus *Conus* (Gastropoda: Toxoglossa) in Hawaii, U.S.A., *Mar. Biol.,* 61, 215, 1981.
12. Christensen, F. B. and Fenchel, T. M., Evolution of marine invertebrate reproductive patterns, *Theor. Popul. Biol.,* 16, 267, 1979.
13. Todd, C. D. and Doyle, R. W., Reproductive strategies of marine benthic invertebrates: a settlement-timing hypothesis, *Mar. Ecol. Prog. Ser.,* 4, 75, 1981.

14. Menge, B. A., Brood or broadcast, the adaptive significance of different reproductive strategies in the two intertidal seastars *Leptasterias hexactis* and *Pisaster ochraceus, Mar. Biol.,* 31, 87, 1975.

15. Todd, C. D., Reproductive energetics of two species of dorid nudibranchs with planktotrophic and lecithotrophic larval strategies, *Mar. Biol.,* 53, 57, 1979.

16. Todd, C. D., The annual cycle of two species of *Onchidoris* (Gastropoda: Opisthobranchia), in *Cyclic Phenomena in Marine Plants and Animals,* Naylor, A. and Hartnell, R., Eds., Pergamon Press, Oxford, 1979, 65.

17. Todd, C. D., The population ecology of *Onchidoris bilamellata* (Gastropoda: Nudibranchia), *J. Exp. Mar. Biol. Ecol.,* 41, 213, 1979.

18. Strathmann, R. R., Egg size, larval development, and juvenile size in benthic marine invertebrates, *Am. Nat.,* 111, 373, 1977.

19. Vance, R. R., Reply to Underwood, *Am. Nat.,* 108, 879, 1974.

20. Strathmann, R. R., The feeding behavior of planktotrophic echinoderm larvae: mechanisms, regulation, and rates of suspension feeding, *J. Exp. Mar. Biol. Ecol.,* 6, 109, 1971.

21. Jagersten, G., *Evolution of the Metazoan Life Cycle: A Comprehensive Theory,* Academic Press, New York, 1972.

22. Thorson, G., Reproductive and larval ecology of marine bottom invertebrates, *Biol. Bull.,* 25, 1, 1950.

23. Strathmann, R. R., The evolution of loss of feeding larval stages in marine invertebrates, *Evolution,* 32, 894, 1978.

24. Rasmussen, E., Faunistic and biological notes on marine invertebrates I., the eggs and larvae of *Brachystomia rissoides* (Hanl.), *Eulimella nitidissima* (Mont.), *Restusa truncatula* (Brug.), and *Embletonia pallida* (Alder and Hancock) (Gastropoda marina), Vidensk, *Medd. Dansk Naturh. Foren.,* 107, 207, 1944.

25. Hadfield, M., Flexibility in larval life history patterns, *Am. Zool.,* 12, 721, 1972.

26. Giglioli, M. E. C., The egg masses of the Naticidae (Gastropoda), *J. Fish. Res. Board Can.,* 12, 287, 1955.

27. Eyster, L. S., Reproduction and developmental variability in the opistobranch *Tenellia pallida, Mar. Biol.,* 51, 133, 1979.

28. Doyle, R. W., Choosing between darkness and light: the ecological genetics of photic behavior in the planktonic larvae of *Spirobis borealis, Mar. Biol.,* 25, 311, 1974.

29. Van Duyl, F. C., Bak, R. M., and Sybesma, J., The ecology of the tropical compound ascidian *Trididemnun solidum,* I. Reproductive strategy and larval behavior, *Mar. Ecol. Prog. Ser.,* 6, 35, 1981.

30. Lesser, J. R., Phyllosoma larvae of *Janus edwardsii* (Hutton) (Crustacea: Palinundae) and their distribution off the east coast of North Island, New Zealand, *N.Z. J. Mar. Fish. Res.,* 12, 357, 1978.

31. Jorgensen, C. B., Feeding and cleaning mechanisms in the suspension feeding bivalve *Mytilus edulis, Mar. Biol.,* 65, 159, 1981.

32. Strathmann, R. R., Progressive vacating of adaptive types during the Phanerozoic, *Evolution,* 32, 907, 1978.

33. Orton, J. H., Sea-temperature, breeding, and distribution of marine animals, *J. Mar. Biol. Assoc. U.K.,* 12, 339, 1920.

34. Kinne, O., Temperature, in *Marine Ecology,* Kinne, O., Ed., Wiley-Interscience, New York, 1970, 407.

35. Sastry, A. N., Physiological adaptation of *Cancer irroratus* larvae to cyclic temperatures, in *Proc. Twelfth Eur. Mar. Biol. Symp.,* McLusky and Berry, A., Eds., Pergamon Press, New York, 1978, 57.

36. Sastry, A. N., Pelecypoda (excluding Ostracidae), in *Reproduction of Marine Invertebrates,* Vol. 5, Giese, A. C. and Pearse, J. S., Eds., Academic Press, New York, 1979, 113.

37. McKenney, C. L. and Neff, J. M., Individual effects and interactions of salinity, temperature, and zinc on larval development of the grass shrimp *Palaemonetes pugio,* I. survival and developmental duration through metamorphosis, *Mar. Biol.,* 52, 177, 1979.

38. Rothlisberg, P. C., Combined effects of temperature and salinity on the survival and growth of the larvae of *Pandalus jordani* (Decopoda: Pandalidae), *Mar. Biol.,* 54, 125, 1979.

39. Christiansen, M. E. and Costlow, J. D., The effect of salinity and cyclic temperature on larval development of the mud-crab *Rhithropanopeus harrisii* (Brachyana: Xanthidae) reared in the laboratory, *Mar. Biol.,* 32, 215, 1975.

40. Johns, D. M., Physiological studies on *Cancer irroratus* larvae, I. Effects of temperature and salinity on survival, developmental rate and size, *Mar. Ecol. Prog. Ser.,* 5, 75, 1981.

41. Johns, D. M., Physiological studies on *Cancer irroratus* larvae, II. Effects of temperature and salinity on physiological performance, *Mar. Ecol. Prog. Ser.,* 6, 309, 1981.

42. Cawthorne, D. F., Tolerance of some cirripede nauplii to fluctuating salinities, *Mar. Biol.*, 46, 321, 1978.

43. Tettlebach, S. T. and Rhodes, E. W., Combined effects of temperature and salinity on embryos and larvae of the Northern Bay scallop *Argopecten irradians*, *Mar. Biol.*, 63, 249, 1981.

44. Costlow, J. D., Jr., Bookhout, C. G., and Monroe, R., Studies on the larval development of the crab *Rhithropanopeus harrisii* Gould, the effects of salinity and temperature on larval development, *Physiol. Zool.*, 39, 81, 1966.

45. Somero, G. N. and Hochachka, P. W., Biochemical adaptations to temperature, in *Adaptation to Environment: Essays on the Physiology of Marine Animals*, Newell, R. C., Ed., Butterworth Publications, London, 1976, 125.

46. Newell, R. C. and Branch, G. M., The influence of temperature on the maintenance of metabolic energy balance in marine invertebrates, *Adv. Mar. Biol.*, 17, 329, 1980.

47. Akesson, B., Temperature and life cycle in *Ophryotrocha labronica* (Polychaeta, Dorvilleidae), *Ophelia*, 15, 37, 1976.

48. Gage, J. D. and Tyler, P. A., Non-viable seasonal settlement of larvae of the upper bathyal brittle star *Ophiocten gracilis* in the Rockwell Trough Abyssal, *Mar. Biol.*, 64, 153, 1981.

49. Sandifer, P. A. and Smith, T. J., Possible significance of variation in the larval development of palaemonid shrimp, *J. Exp. Mar. Biol. Ecol.*, 39, 55, 1979.

50. Knowlton, R. E., Larval developmental processes and controlling factors in decapod crustacea, with emphasis on Caridea, *Thalassia Jugosl.*, 10, 138, 1974.

51. Lebour, M. V., The food of plankton organisms II, *J. Mar. Biol. Assoc. U.K.*, 13, 70, 1923.

52. Lucas, J. S., Hart, R. J., Howden, M. E., and Salathe, R., Saponins in eggs and larvae of *Acanthaster planci* (L.) (Asteroidea) as chemical defenses against planktivorous fish, *J. Exp. Mar. Biol. Ecol.*, 40, 155, 1979.

53. Burrell, V. C. and Van Engel, V. A., Predation by and distribution of a ctenophore, *Mnemiopsis leidyi* Agassiz, in New York estuary, *Estuarine Coastal Mar. Sci.*, 4, 235, 1976.

54. Jorgensen, C. B., *Biology of Suspension Feeding*, Pergamon Press, Oxford, 1966.

55. Blake, J., Reproduction and larval development of *Polydora* from Northern New England (Polychaete: Spionidae), *Ophelia*, 7, 1, 1969.

56. Garstang, W., The origin and evolution of larval forms, Presidential address, section D. *Br. Assoc. Adv. Sci.*, Glasgow, 1928, 22.

57. Forward, R. B., Crustacean larval negative phototaxis: possible functional significance, *J. Exp. Mar. Biol. Ecol.*, 16, 11, 1974.

58. Forward, R. B., Light and diurnal vertical migration: photo behavior photophysiology of plankton, in *Photochemical and Photobiological Reviews*, Vol. 1, Smith, K., Ed., Plenum Press, New York, 1976, 157.

59. Forward, R. B., The occurrence of a shadow response among brachyuran larvae, *Mar. Biol.*, 39, 331, 1977.

60. Pechenik, J. A., Role of encapsulation in invertebrate life histories, *Am. Nat.*, 114, 859, 1979.

61. Caswell, H., The evolution of mixed life histories in marine invertebrates and elsewhere, *Am. Nat.*, 117, 529, 1981.

62. Crisp, D. J., Settlement responses in marine organisms, in *Adaptation to Environment: Essays on the Physiology of Marine Animals*, Newell, R. C., Butterworth Publications, London, 1976, 83.

63. Holland, D. L. and Walker, G., The biochemical composition of the cypris larva of the barnacle *Balanus balanoides*, *J. Cons. Int. Explor. Mer.*, 36, 162, 1975.

64. Lucas, M. I., Walker, G., Holland, D. L., and Crisp, D. J., An energy budget for the free-swimming and metamorphosing larvae of *Balanus balanoides* (Crustacea: Cirripedia), *Mar. Biol.*, 55, 221, 1979.

65. Millar, R. H. and Scott, J. M., The larvae of the oyster *Ostraea edulis* during starvation, *J. Mar. Biol. Assoc. U.K.*, 47, 475, 1967.

66. Bayne, B. L., Growth and the delay of metamorphosis of the larvae of *Mytilus edulis*, L. *Ophelia*, 2, 1, 1965.

67. Paul, A. J., Paul, J. M., Shoemaker, P. A., and Feder, H. M., Prey concentration and feeding response in laboratory-reared stage one zoeae of king crab, snow crab, and pink shrimp, *Trans. Am. Fish. Soc.*, 108, 440, 1979.

68. Paul, A. J. and Paul, J. M., The effect of early starvation on later feeding success of king crab zoeae, *J. Exp. Mar. Biol. Ecol.*, 44, 247, 1980.

69. Bigford, T. E., Effect of several diets on survival, development time, and growth of laboratory-reared spider crab *Libinia emarginata*, larvae, *Fish. Bull. U.S.A.*, 76, 59, 1978.

70. Anger, K. and Dawirs, R. R., Influence of starvation on the larval development of *Huas araneus* (Decapoda: Majidae), *Helgol. Wiss. Meeresunter.*, 34, 287, 1981.

71. Sulkin, S. D., The significance of diet in the growth and development of larvae of the blue crab *Callinectes sapidus* under laboratory conditions, *J. Exp. Mar. Biol. Ecol.*, 20, 119, 1975.

72. Sulkin, S. D. and Norman, K., A comparison of two diets in the laboratory culture of the zoeal stages of the brachyuran crabs *Rhithropanopeus harrisii* and *Neopanope* sp., *Helgol. Wiss. Meeresunter.*, 28, 183, 1976.

73. Sulkin, S. D., Nutritional requirements during larval development of the portunid crab *Callinectes sapidus* Rathkin, *J. Exp. Mar. Biol. Ecol.*, 34, 29, 1978.

74. Jones, D. A., Kanazawa, A., and Ono, K., Studies on the nutritional requirements of the larval stages of *Penaeus japonicus* using microencapsulated diets, *Mar. Biol.*, 54, 261, 1979.

75. Sastry, A. N., The development and external morphology of pelagic larval and post larval stages of the bay scallop, *Aequipecten irradians concentricus,* reared in the laboratory, *Bull.. Mar. Sci. Gulf Caribb.*, 15, 417. 1965.

76. Day, R., Reproduction, development, and larval energetics of the spionid polychaete *Polydora websteri*, Ph.D. dissertation, University of South Florida, Tampa, 1982, p. 109.

77. Anger, K., Dawirs, R. R., Anger, J., and Costlow, J. D., Effects of early starvation period on zoeae development of Brachyuran crabs, *Biol. Bull.*, 161, 199, 1981.

78. Holland, D. L., Lipid reserves and energy metabolism in the larvae of benthic marine invertebrates, *Biochem. Biophys. Pers. Mar. Biol.*, 4, 85, 1978.

79. Isono, Y. and Isono, N., Biochemistry and Morphogenesis, in *The Sea Urchin Embryo,* Czihak, G., Ed., Springer Verlag, Berlin, 1975, 608.

80. Yanagisawa, T., Respiration and energy metabolism, in *The Sea Urchin Embryo,* Czihak, G., Ed., Springer Verlag, Berlin, 1975, 510.

81. Yanagisawa, T., Carbohydrate metabolism and related enzymes, in *The Sea Urchin Embryo,* Czihak, G., Ed., Springer Verlag, Berlin, 1975, 550.

82. Holland, D. L. and Hannant, P. J., Addendum to a microanalytical scheme for the biochemical analysis of marine invertebrate larvae, *J. Mar. Biol. Assoc. U.K.*, 53, 833, 1973.

83. Needham, J., *Biochemistry and Morphogenesis,* Cambridge University Press, Cambridge, 1950.

84. Barnes, H., Studies in the biochemistry of cirriped eggs, I. Changes in general development of *Balanus balanoides* and *Balanus balanus,* *J. Mar. Biol. Assoc. U.K.*, 45, 321, 1965.

85. Holland, D. L. and Hannant, P. J., The glycogen content in winter and summer of oysters, *Ostrea edulis* of different ages, *J. Cons. Cons. Int. Explor. Mer.*, 36, 240, 1976.

86. Pandian, T. J., Yolk utilization in the gastropod *Crepidula tornicata, Mar. Biol.*, 3, 117, 1969.

87. Dawson, M. C. and Barnes, H., Studies in the biochemistry of cirriped eggs, II. Changes in lipid composition during development of *Balanus balanoides* and *Balanus balanus,* *J. Mar. Biol. Assoc. U.K.*, 46, 249, 1966.

88. Bayne, B. L., Gabbott, P. A., and Widdows, J., Some effects of stress in the adult on the eggs and larvae of *Mytilus edulis, L., J. Mar. Biol. Assoc. U.K.*, 55, 675, 1975.

89. Millar, R. H. and Scott, J. M., The larva of the oyster *Ostrea edulis* during starvation, *J. Mar, Biol. Assoc. U.K.*, 47, 475, 1967.

90. Holland, D. L. and Spencer, B. E., Biochemical changes in fed and starved oyster, *Ostrea edulis* L. during larval development, metamorphosis and early spat growth, *J. Mar. Biol. Assoc. U.K.*, 53, 287, 1973.

91. Helm, M. M., Holland, D. L., and Stevenson, R. R., The effect of supplementary algal feeding of a hatchery breeding stock *Ostrea edulis* L. on larval vigour, *J. Mar. Biol. Assoc. U.K.*, 53, 673, 1973.

92. Isono, N. and Yasumasu, I., Pathways of carbohydrate breakdown in sea urchin eggs, *Exp. Cell Res.*, 50, 616, 1968.

93. Frank, J. R., Sulkin, S. D., and Morgan, R. P., Biochemical changes during larval development of the xanthid crab *Rhithropanopeus harrisii* I. Protein, total lipid, alkaline phosphatase and glutamic oxaloacetic transaminase, *Mar. Biol.*, 32, 105, 1975.

94. Sulkin, S. D., Morgan, R. P., and Minasian, L. L., Biochemical changes during larval development of the xanthid crab *Rhithropanopeus harrisii* II. Nucleic acids, *Mar. Biol.*, 32, 113, 1975.

95. Whitney, J. O., Absence of sterol synthesis in larvae of the mud-crab *Rhithropanopeus harrisii* and of the spider crab *Libinia emarginata, Mar. Biol.*, 3, 134, 1969.

96. Achituv, Y., Blackstock, J., Barnes, M., and Barnes H., Some biochemical constituents of stage I and II nauplii of *Balanus balanoides* (L.) and the effect of anoxia on stage I, *J. Exp. Mar. Biol. Ecol.*, 42, 1, 1980.

97. Rainbow, P. J. and Walker, G., The functional morphology and development of the alimentary tract of larval and juvenile barnacles (Cirripedia: Thoracia), *Mar. Biol.*, 42, 337, 1977.

98. West, T. L. and Costlow, J. D., Molting control in sixth stage nauplii of *Balanus eburneus, Am. Zool.*, 20, 843, 1980.

99. Davenport, J., A technique for the measurement of oxygen consumption in small aquatic organisms, *Lab. Pract.*, 25, 693, 1976.

100. Lawrence, J. M., Patterns of lipid storage in post-metamorphic marine invertebrates, *Am. Zool.*, 16, 747, 1976.

101. Sargent, J. R., The structure, metabolism, and function of lipids in marine organisms, *Biochem. Biophys. Pers. Mar. Biol.*, 3, 149, 1976.

102. Holland, D. L., Tawtamasiriwong, R., and Hannant, P. J., Biochemical composition and energy reserves in the larvae and adults of the four British periwinkles *Littorina littorea, L. littoralis, L. saxatilis, L. neritoides, Mar. Biol.*, 33, 235, 1975.

103. Waldock, M. J. and Holland, D. L., Fatty acid composition of the triacylglycerols of the cypris larvae of the barnacle *Balanus balanoides* during metamorphosis, *Mar. Biol.*, 46, 163, 1978.

104. Kempf, S., Acquisition, storage, and utilization of nutrients by the embryos and larvae of opistobranch molluscs, Ph.D. dissertation, University of Honolulu, Hawaii, 1982.

105. Bonar, D. B. and Hadfield, M., Metamorphosis of the marine gastropod *Phestilla sibogae* (Nudibranchia: Aeolidacea), I. light and electron microscopic analysis of larval and metamorphic stages, *J. Exp. Mar. Biol. Ecol.*, 16, 227, 1974.

106. Bonar, D. B., Ultrastructure of a cephalic sensory organ in larvae of the gastropod *Phestilla sibogae* (Nudibranchia: Acolidacea), *Tissue and Cell*, 10, 153, 1978.

107. Wilson, D. P., The influence of the nature of the substratum on the metamorphosis of the larvae of marine animals, especially the larvae of *Ophilia bicornis* Savigny, *Ann. Inst. Oceanogr.*, Monaco, 27, 49, 1952.

108. Scheltema, R. S., Metamorphosis of the veliger larvae of *Nassarius obsoletus* (Gastropods) in response to bottom sediment, *Biol. Bull.*, 120, 92, 1961.

109. Dean, D. and Blake, J. A., Life history of *Boccardia hamata* (Webster) on the east and west coasts of North America, *Biol. Bull.*, 130, 316, 1966.

110. Crisp, D. J., Factors influencing the settlement of marine invertebrate larvae, in *Chemoreception in Marine Organisms*, Grant, P. T. and Mackie, D. M., Eds., Academic Press, New York, 1974, 177.

111. Switzer-Dunlap, M. and Hadfield, M. G., Reproductive patterns of Hawaiian aplysiid gastropods, in *Reproductive Ecology of Marine Invertebrates*, Stancyk, S. E., Ed., University of South Carolina Press, Columbia, South Carolina, 1977.

112. Day, R. L. and Blake, J. A., Reproduction and larval development of *Polydora giardi* Mesnil (Polychaeta: Spionidae), *Biol. Bull.*, 156, 20, 1979.

113. Scheltema, R. S., Evidence for trans-atlantic transport of gastropod larvae belonging to the genus *Cymatium, Deep-Sea Res.*, 13, 83, 1966.

114. Scheltema, R. S., The dispersal of the larvae of shoalwater benthic invertebrate species over long distances by ocean currents, in *Proc. Fourth Eur. Mar. Biol. Symp.*, Crisp, D. J., Ed., Cambridge University Press, Cambridge, 1971, 7.

115. Vernberg, F. J., Dissolved gases, in *Marine Ecology*, Kinne, O., Ed., Wiley-Interscience, New York, 1972, 1491.

116. Hinegardiner, R. T., Growth and development of the laboratory cultured sea urchin, *Biol. Bull.*, 137, 465, 1969.

117. Kempf, S. C., Long-lived larvae of the gastropod *Aplysia juliana*: do they disperse and metamorphose or just slowly fade away, *Mar. Ecol. Prog. Ser.*, 6, 61, 1981.

118. Pechenik, J. A., Growth and energy balance during the larval lives of three prosobranch gastropods, *J. Exp. Mar. Biol. Ecol.*, 44, 1, 1980.

119. Knight-Jones, E. W., Gregariousness and some other aspects of the settling behavior of *Spirorbis, J. Mar. Biol. Assoc. U.K.*, 30, 201, 1953.

120. Williams, G. B., The effects of extracts of *Fucus serratus* in promoting settlement of larvae of *Spirobis borealis* (Polychaeta), *J. Mar. Biol. Assoc. U.K.*, 44, 397, 1964.

121. Loosonoff, V. L., New advances in the study of bivalve larvae, *Am. Sci.*, 42, 607, 1954.

122. Reeve, M. R., Growth, metamorphosis, and energy conversion in the larva of the prawn, *Palaemon serratus, J. Mar. Biol. Assoc. U.K.*, 49, 77, 1969.

123. Day, R. L., Studies on the reproduction and larval biology of *Polydora giardi* Mesnil (Polychaeta: Spionidae), Master Thesis, University of the Pacific, Stockton, Calif., 1977.

124. Shelbourne, J. E., The feeding and condition of plaice larvae in good and bad plankton patches, *J. Mar. Biol. Assoc. U.K.*, 36, 539, 1957.

125. Lasker, R., Field criteria for survival of anchovy larvae: the relation between inshore chlorophyll maximum layers and successful first feeding, *Fish. Bull., U.S.*, 73, 453, 1975.

126. Wibory, K. F., Larval mortality in marine fishes and the critical period concept, *J. Cons. Int. Explor. Mar.*, 37, 11, 1976.

127. Scheltema, R., On the relationship between dispersal of pelagic veliger larvae and the evolution of marine prosobranch gastropods, in *Marine Organisms Genetics, Ecology and Evolution*, Battaglia, B. and Beardmore J., Eds., Plenum Press, New York, 1978, 303.

128. Scheltema, R., Dispersal of marine invertebrate organisms: Paleobiogeographic and biostratigraphic implications, in *Concepts and Methods of Biostratigraphy*, Kauffman, E. G. and Hazel, J. E., Eds., Dowden, Hutchinson, and Ross, Pennsylvania, 1977, 73.

129. Thorson, G., Length of pelagic larval life in the marine bottom invertebrates as related to larval transport by ocean currents, in *Oceanography*, Sears, M., Ed., American Association for the Advancement of Science, Washington, D.C., 1961, 455.

130. Crisp, D. J., Genetic consequences of different reproductive strategies in marine invertebrates, in *Marine Organisms Genetics, Ecology and Evolution*, Battaglia, B. and Beardmore, J., Eds., Plenum Press, New York, 1978, 257.

131. Crisp, D. J., The spread of *Elminuis modestus* Darwin in northwest Europe, *J. Mar. Biol. Assoc. U.K.*, 37, 483, 1958.

132. Quayle, D. B., Distribution of introduced marine Mollusca in British Columbia waters, *J. Fish. Res. Board Canada*, 21, 1155, 1964.

133. Gerdes, D., The re-establishment of an *Amphiura filiforms* (Miller) population in the inner part of the German bight, in *Biology of Benthic Organisms*, Keegan, B. F., O'Ceidigh, P., and Boaden, P. J. S., Eds., Pergamon Press, New York, 1977, 277.

134. Okubo, A., Oceanic diffusion diagrams, *Deep-Sea Res.*, 18, 789, 1971.

135. Strathmann, R. R., The spread of sibling larvae of sedentary marine invertebrates, *Am. Nat.*, 108, 29, 1974.

136. Gardiner, D., Notes and observations on the distribution of the larvae of marine animals, *Ann. Mag. Nat. Hist. Ser.*, 7, 403, 1904.

137. Ekman, S., *Zoogeography of the Sea*, Sedgewick and Jackson, Ltd., London, 1953.

138. Lockwood, A. P. M., Physiological adaptation to life in estuaries, in *Adaptation to Environment: Essays on the Physiology of Marine Animals*, Newell, R. C., Ed., Butterworth, London, 1976, 375.

139. deWolf, P., On the retention of marine larvae in estuaries, *Thalassia Jugosl.*, 10, 415, 1974.

140. Wood, L. and Hargis, W. J., Jr., Transport of bivalve larvae in a tidal estuary, in *Fourth Eur. Mar. Biol. Symp.*, Crisp, D. J., Ed., Cambridge University Press, New York, 1971, 29.

141. Palmer, A. R. and Strathmann, R. R., Scale of dispersal in varying environments and its implications for life histories of marine invertebrates, *Oecologia*, 48, 308, 1981.

142. Kuno, E., Dispersal and the persistence of populations in unstable habitats, a theoretical note, *Oecologia*, 49, 123, 1981.

143. Cohen, D., Optimizing reproduction in a randomly varying environment, *J. Theor. Biol.*, 12, 119, 1966.

144. Gadgil, M., Dispersal consequences and evolution, *Ecology*, 52, 253, 1971.

145. Roff, D. A., An analysis of a population model demonstrating the importance of dispersal in a heterogeneous environment, *Oecologia*, 15, 259, 1974.

146. Sandifer, P. A., The role of the pelagic larvae in recruitment to populations of adult crustaceans in the York River Estuary and adjacent lower Chesapeake Bay, Virginia, *Estuarine Coastal Mar. Sci.*, 3, 269, 1975.

147. Sandifer, P. A., Distribution and abundance of decapod larvae in the York River Estuary and adjacent lower Chesapeake Bay, Virginia, *Chesapeake Sci.*, 14, 235, 1973.

148. Bousfield, E. L., Ecological control of occurrence of barnacles in the Miramichi estuary, *Bull. Natn. Mus. Can.*, 137, 69, 1955.

149. Sulkin, S. D. and Van Heukelem, W. F., Ecological and evolutionary significance of nutritional flexibility in planktotrophic larvae of the deep sea red crab *Geryon quinquedens* and the stone crab *Menippe mercenaria*, *Mar. Ecol. Prog. Ser.*, 2, 91, 1980.

150. deWolf, P., Ecological observations on the mechanisms of dispersal of barnacle larvae during planktonic life and settling, *Neth. J. Sea. Res.*, 6, 1, 1973.

151. Thorson, G., Light as an ecological factor in the dispersal and settlement of larvae of marine bottom invertebrates, *Ophelia*, 1, 167, 1964.

152. Crisp, D. J. and Ritz, D. A., 1973, Responses of cirripede larvae to light, I. Experiments with white light, *Mar. Biol.*, 23, 327, 1973.

153. Sulkin, S. D., Depth regulation of crab larvae in the absence of light, *J. Exp. Mar. Biol. Ecol.*, 13, 73, 1973.

154. Latz, M. I. and Forward, R. B., The effect of salinity upon phototaxis and geotaxis in a larval crustacean, *Biol. Bull.*, 153, 163, 1977.

155. Bentley, E. and Sulkin, S. D., The ontogeny of barokinesis during zoeal development of the xanthid crab *Rhithropanopeus harrisii* (Gould), *Mar. Behav. Physiol.*, 4, 275, 1977.

156. Wheeler, D. E. and Epifanio, C. E., Behavioral responses to hydrostatic pressure in larvae of two species of xanthid crabs, *Mar. Biol.*, 46, 167, 1978.

157. Carriker, M. R., Ecological observations on the distribution of oyster larvae in New Jersey estuaries, *Ecol. Monogr.*, 21, 19, 1951.

158. Kalber, F. A. and Costlow, J. D., The ontogeny of osmoregulation and its neurosecretory control in decapod crustacean *Rhithropanopeus harrisii* (Gould), *Am. Zool.*, 6, 221, 1966.

119

159. Cronin, T. W. and Forward, R. B., Tidal vertical migration: an endogenous rhythm in estuarine Crab larvae, *Science*, 205, 1020, 1979.
160. Forward, R. B. and Costlow, J. D., The ontogeny of phototaxis by larvae of the crab *Rhithropanopeus harrisii* (Gould), *Mar. Biol.*, 26, 27, 1974.
161. Forward, R. B. and Cronin, T. W., Spectral sensitivity of larvae from intertidal crustaceans, *J. Comp. Physiol.*, 133, 311, 1979.
162. Via, S. E. and Forward, R. S., The ontogeny and spectral sensitivity of polarataxis in larvae of the crab *Rhithropanopeus harrisii*, *Biol. Bull.*, 149, 251, 1975.
163. Ott, F. S. and Forward, R. B., The effect of temperature on phototaxis and geotaxis by larvae of the crab *Rhithropanopeus harrisii*, *J. Exp. Mar. Biol. Ecol.*, 23, 97, 1976.
164. Cronin, T. W. and Forward, R. B., The effects of starvation on phototaxis and swimming of larvae of the crab *Rhithropanopeus harrisii*, *Biol. Bull.*, 158, 283, 1980.
165. Wooldridge, T. and Erasmus, T., Utilization of tidal currents by estuarine zooplankton, *Estuarine Coastal Mar. Sci.*, 11, 107, 1980.
166. Sulkin, S. D. and Van Heukelem, W., Larval recruitment in the crab *Callinectes sapidus* Rathbun: An amendment to the concept of larval retention in estuaries, in *Estuarine Comparisons*, Kennedy, U. S., Ed., Academic Press, New York, in press.
167. Scheltema, R., Biological interactions determining larval settlement of marine invertebrates, *Thalassia Jugosl.*, 10, 263, 1974.
168. Chia, F. S. and Rice, M. E., Eds., *Settlement and Metamorphosis of Marine Invertebrate Larvae*, Elsevier, New York, 1978.
169. Young, C. M., and Braithwaite, L. F., Larval behavior and post settling morphology in the ascidian *Chelyosoma productum*, *J. Exp. Mar. Biol. Ecol.*, 42, 152, 1980.
170. Birkeland, C., Chia, F. S., and Strathmann, R., Development, substratum selection, delay of metamorphosis and growth in the sea star, *Mediaster aequalis* Stimpson, *Biol. Bull.*, 141, 99, 1971.
171. Hadfield, M. G., Personal Communication, 1982.
172. Obreski, S., Evolution of larval colonizing strategy, *Mar. Ecol. Prog. Ser.*, 1, 293, 1980.
173. Bonar, D. B., Regeneration of larval and potential adult tissues in sea-urchin larvae, *Am. Zool.*, 19, 926, 1979.
174. Bonar, D. B., Regeneration in sea-urchin larvae, *Am. Zool.*, 18, 581, 1978.
175. Sulkin, S. D., Van Heukelem, W., and Van Heukelem, L., The behavioral basis of larval recruitment in the crab *Callinectes sapidus* Rathbun: A laboratory investigation of otogenetic changes in geotaxis and barokinesis, *Biol. Bull.*, 159, 402, 1980.
176. Kriegstein, A. R., Castellucci, V., and Kandel, E. R., Metamorphosis of *Aplysia californica* in laboratory culture, *Proc. Nat. Acad. Sci. U.S.A.*, 71, 3654, 1974.
177. Jackson, G. A. and Strathmann, R. R., Larval mortality from offshore mixing as a link between precompetent and competent periods of development, *Am. Nat.*, 118, 16, 1981.
178. Doyle, R. W., Settlement of planktonic larvae: A theory in habitat selection in varying environments, *Am. Nat.*, 109, 113, 1973.
179. Scheltema, R. S., Dispersal of larvae as a means of genetic exchange between widely separated populations of schoalwater benthic invertebrates species, in *Fifth Eur. Mar. Biol. Symp.*, Battaglia, E., Ed., Piccin, Padua, Italy, 1972, 101.
180. Scheltema, R. S., Trans-Atlantic dispersal of larvae from benthic organisms by means of ocean currents, in *Tech. Pap., Woods Hole Oceanogr. Inst., Abstr.*, 196J, 1965.
181. Scheltema, R. S., Dispersal of larvae by equatorial ocean currents and its importance to the zoogeography of shoalwater tropical species, *Nature*, 217, 1159, 1968.
182. Heck, K. L. and McCoy, E. D., Long-distance dispersal and the reef-building corals of the Eastern Pacific, *Mar. Biol.*, 48, 349, 1978.
183. Stearns, S. C., Life-History tactics: A review of the ideas, *Q. Rev. Biol.*, 51, 3, 1976.
184. MacArthur, R. H. and Wilson, E. O., *The Theory of Island Biogeography*, Princeton University Press, Princeton, New Jersey, 1967.
185. Williams, G. C., *Sex and Evolution*, Princeton University Press, Princeton, New Jersey, 1975.
186. Scheltema, R. S., Dispersal of pelagic larvae and the zoogeography of tertiary benthic gastropods, in *Historical Biogeography, Plate Tectonics and the Changing Environment*, Boucot, A. J. and Gray, J., Eds., Oregon State University Press, Corvallis, Oregon, 1979.
187. Dana, T. B., Development of contemporary Eastern Pacific coral reefs, *Mar. Biol.*, 33, 355, 1975.
188. Coe, W. R., Resurgent populations of the littoral marine invertebrates and their dependence on ocean currents, *Ecology*, 34, 225, 1953.
189. Knight-Jones, E. W. and Moyse, J., Intraspecific competition in sedentary marine animals, *Symp. Soc. Exp. Biol.*, 15, 72, 1961.
190. Scheltema, R. S., Williams, I. P., Shaw, M. A., and Loudon, C., Gregarious settlement by the larvae of *Hydroides dianthus* (Polychaeta: Serpulidae) *Mar. Ecol. Prog. Ser.*, 5, 69, 1981.

191. Moyse, J. and Hui, E., Avoidance by *Balanus balanoides* cyprids of settlement on conspecific adults, *J. Mar. Biol. Assoc. U.K.,* 61, 449, 1981.
192. Fauchald, K., Personal Communication, 1982.
193. Grassle, J. F. and Grassle, J. P., Life histories and genetic variation in marine invertebrates, in *Marine Organisms Genetics, Ecology, and Evolution,* Battaglia, B. and Beardmore, J. A., Eds., Plenum, New York, 1978, 347.
194. Strathmann, R. R., Branscomb, E. S., and Vedder, K., Fatal errors in set as a cost of dispersal and the influence of intertidal flora on set of barnacles, *Oecologia,* 48, 13, 1981.
195. Shuto, T., Larval ecology of prosobranch gastropods and its bearing on biogeography and paleontology, *Lethaia,* 7, 239, 1974.
196. Jablonski, D., Apparent versus real biotic effects of transgressions and regressions, *Paleobiology,* 6, 397, 1980.
197. Jablonski, D. and Lutz, D. A., Molluscan larval shell morphology: ecological and paleontological applications, in *Skeletal Growth of Aquatic Organisms,* Rhodes, D. C. and Lutz, R. A., Eds., Plenum Press, New York, 1980.
198. Hansen, T. A., Larval dispersal and species longevity in lower tertiary gastropods, *Science,* 199, 885, 1978.
199. Hansen, T. A., Influence of larval dispersal and geographic distribution on species longevity in neo-gastropods, *Paleobiology,* 6, 193, 1980.
200. Levins, R., The theory of fitness in a heterogeneous environment. IV. The adaptive significance of gene flow, *Evolution,* 17, 635, 1964.
201. Lewontin, R. C., *The Genetic Basis of Evolutionary Change,* Columbia University Press, New York, 1974.
202. Grassle, J. F., Species diversity, genetic diversity, and environmental uncertainty in *Fifth Eur. Mar. Biol. Symp.,* Battaglia, B., Ed., Piccin, Padua, Italy, 1972, 19.
203. Gooch, J. L. and Schopf, T. J. M., Genetic variability in the deep sea: Relation to environmental uncertainty, *Evolution,* 26, 545, 1972.
204. Valentine, J. W. and Ayala, F. J., Genetic variability in krill, in *Proc. Nat. Acad. Sci. U.S.A.,* 1976, 658.
205. Ayala, F. J., DeLaca, T. E., and Zumwalt, G. S., Genetic variability of the Antarctic brachiopod *Liothyrella notorcadensis* and its bearing on mass extinction hypothesis, *J. Paleontol.,* 49, 1, 1975.
206. Valentine, J. W., Genetic strategies of adaptation, in *Molecular Evolution,* Ayala, F. J., Ed., Sinauer, Sunderland, Mass., 1976, 78.
207. Gooch, J. L., Smith, B. S., and Knupp, D., Regional survey of gene frequencies in the mud snail *Nassarius obsoletus, Biol. Bull.,* 142, 36, 1972.
208. Koehn, R. K., Migration and population structure in the pelagically dispersing marine invertebrate *Mytilus edulis,* in *Isozymes, N. Genetics and Evolution,* Markert, C. L., Ed., Academic Press, New York, 1975, 945.

Chapter 6

LIFE HISTORY AND ECOLOGY OF PELAGIC FISH EGGS AND LARVAE

Mark M. Leiby

TABLE OF CONTENTS

I. INTRODUCTION

The majority of estuarine and marine fishes, whether the adults are benthic, bentho-pelagic, mesopelagic, or pelagic, have eggs and/or larvae which are members of the estuarine or marine planktonic community. The types of reproduction, the means of introducing the eggs or larvae to the planktonic community, and the methods of ensuring dispersal to new areas or recruitment into an existing population which are used by adult fishes are nearly as numerous as are the families of fishes.

It would be impossible to detail all the various combinations of, and gradations between, the methods of reproduction, egg and larval broadcast, dispersal and recruitment (and the limiting factors involved in each) which are used by fish, especially since the early life history of many fishes is known only in general terms or not at all. Consequently, this paper will be an overview of: (1) methods of reproduction used by fish; (2) methods of egg or larval introduction into the planktonic community; (3) egg and larval dispersal; and (4) limiting factors affecting the survival and subsequent recruitment of larval fishes.

II. METHODS OF REPRODUCTION

In order for a species to avoid extinction it must reproduce at a level which is sufficient to insure recruitment into the existing stock of enough young to maintain the stock at or above some minimum critical level. Fish meet this requirement in a variety of ways. At one end of the scale, some fish put the energy resources available for reproduction into the production of relatively few eggs and young (less than 50 in the surf perch *Amphistichus rhodopterus*[1]) which are protected in some manner by the parent(s) during the most vulnerable stages of development, thus insuring a high percentage of survival. At the other end of the scale, some fishes put their available energy resources into the production of a large number of eggs (reportedly as many as 12,000,000 in the tarpon *Megalops atlanticus*[2]) which are left to the mercy of the seas and which have a very low, albeit sufficient, percentage of survival.

The range of reproductive mechanisms used by fish is generally broken down into three categories; viviparous, ovoviviparous and oviparous reproduction. Viviparous fish produce eggs which hatch inside the female with the developing young meeting nutritional, excretory and respiratory needs by means of a connection with maternal tissues, although not necessarily with the females circulatory system. Ovoviviparous fish produce eggs which, although they hatch inside the female, have sufficient yolk present in the egg to meet nutritional needs of the developing young and the female provides only protection during the most vulnerable stages of development. Oviparous fish produce eggs which hatch outside the female and which, depending on the species, may or may not be protected by one or both parents. In reality, these categories are categories of convenience for discussion since there is very nearly a continuum in modes of reproduction and development from viviparous to oviparous. If one were to list all the methods for fertilization of the eggs and development of the young which fish are known to use, just about the only combination of these methods which has not been found would be externally fertilized eggs which are reintroduced into the female's reproductive system for development. Among the fishes using most of these combinations, there are species which produce eggs and/or larvae which, if only for a brief period, are part of the planktonic community.

A. Viviparous Fishes

Viviparity, which has evolved independently in various fish orders and families,[3,4] is found in the elasmobranchs and in the teleost orders Atheriniformes, Ophidiiformes, Gadiformes, and Perciformes.[3-8] While elasmobranchs have many interesting adaptations to viviparity, including the development of an umbilical attachment to a placenta[9] and intrauterine canabalism,[10-12] the young at birth are too large to be considered members of the planktonic community and will not be dealt with here. Viviparous teleosts are not known to have an umbilical attachment to a placental-like structure, but have derived a variety of intriguing mechanisms for obtaining their needs from maternal tissues.[3,5,13-30] These references will provide the interested reader with a thorough discussion of these adaptations. Most viviparous teleosts produce young which are in an advanced state of development at birth and are not part of the planktonic community.[28] Some species, however, do produce young which are part of the planktonic community, if only for a short time.[5,6,31,32]

B. Ovoviviparous Fishes

Ovoviviparity, which has also evolved independently in various fish families, and which is the logical evolutionary predecessor to viviparity,[4,32] is found in the elasmobranchs, Coelacanthiformes, and the teleost orders Atheriniformes, Ophidiiformes, Gadiformes, Scorpaeniformes, and Perciformes.[1,3,4,7,10,32-34] The difference between viviparity and ovoviviparity in fishes is not distinct. In some species initial development of the embryo depends on the yolk supply invested in the egg with the developing young becoming viviparous later during growth.[3,4,35] Many embryos which do not have any apparent morphological adaptations for utilization of maternal resources derive at least some nourishment from the mother sometime during their development.[3,4,34] Only a few species such as *Sebastodes paucispinis* and *Dinematichthys ilucoeteoides* are unequivocally ovoviviparous.[32,33] Not all ovoviviparous species have young which are planktonic, but, in general, the incidence of planktonic larvae or juveniles seems to be greater in ovoviviparous fish than in viviparous fish.

The larvae and juveniles of viviparous and ovoviviparous fish which are planktonic are relatively large and in an advanced state of development when they enter the planktonic community. Consequently, they bypass the "critical period" which is so limiting for fish with pelagic eggs, and their survival rate is quite high. However, since egg production by viviparous and ovoviviparous fish is relatively low compared to oviparous fish, and since those larvae which are planktonic generally spend a shorter time in the plankton than do the eggs and larvae of oviparous fish, dispersal of the species and colonization of new areas or recolonization of depleted areas is liable to be slower and less widespread than it is for fishes with planktonic eggs and larvae, particularly for those viviparous and ovoviviparous fish which are bottom dwellers as adults. The planktonic larvae and juveniles of viviparous and ovoviviparous fish are only a small portion of the ichthyoplankton, and are subject to the same forces affecting the later stages of the planktonic larvae of oviparous fish.

C. Oviparous Fishes

Oviparity has been reported, or is suspected, in all orders of gnathostomous fishes except in a few orders of the elasmobranchs and in the Coelacanthiformes. It is by far the most common reproductive mechanism found in fishes.

A variety of successful strategies for incubation of fertilized eggs have evolved in estuarine and marine oviparous fish. These strategies generally fall into the following four categories: (1) massed eggs which receive parental care; (2) massed eggs which do not receive parental care; (3) single eggs which receive parental care; (4) single eggs which do not receive parental care. Some, but not all, species using each of these strategies produce eggs and/or larvae which are part of the planktonic community.

Examples of each strategy and some of the variations found within each strategy are as follows.

Massed eggs receiving parental care — In marine fishes this strategy is known in the orders Scorpaeniformes, Perciformes, and Tetraodontiformes. The lumpfish *Cyclopterus lumpus* deposits large clusters of eggs (up to 200,000) on rocks. These egg masses are protected by the males until hatching.[7] The young are planktonic, frequently being associated with drifting seaweed.[35,36] The prickleback *Opisthocentrus ocellatus* lays an egg mass in narrow cavities under stones with the female guarding them.[37] The triggerfish *Pseudobalistes flavimarginatus* and *Balistopus undulatus* lay spongy clusters of eggs in a depression which has been created by the adult. The eggs are guarded by the adults until hatching when they become planktonic.[38]

Massed eggs not receiving parental care — In marine fishes this strategy is known in the orders Lophiiformes, Scorpaeniformes, and Ophidiiformes. The goosefish *Lophius americanus* produces free-floating mucoid egg rafts or veils which are in single layers and which may be as much as nine meters long and one meter wide and which may contain more than 1,000,000 eggs.[7,39,40] The sargassum fish *Histrio histrio* also produces free-floating rafts of eggs although they are smaller than the rafts produced by *Lophius*.[7] The scorpionfish *Scorpaena guttata* produces a free-floating, thin, gelatinous matrix of eggs.[7,41] The cusk eel *Ophidium barbatum* produces eggs which are found floating in gelatinous masses,[7] and the same condition has been reported in the pearlfish *Carapus imberbis*.[7] All of these species produce planktonic larvae.

Single eggs receiving parental care — This strategy, which takes a variety of forms, is known to produce planktonic larvae and juveniles in the marine and estuarine orders Lophiiformes, Syngnathiformes, Gasterosteiformes, Perciformes, Gobiesociformes, and Tetraodontiformes.

The anglerfish *Antennarius caudimaculatus* is known to carry the eggs attached to the side of the male.[40] In the family Syngnathidae the male pipefish and seahorses incubate the eggs in brood pouches located under the abdomen or under the tail.[7] In the related family Solenostomidae, the false pipefishes, the females have the brood pouch. The tubesnout fish *Aulorhynchus flavidus* builds nests in kelp beds with the male guarding the nests.[42,43] In the order Perciformes, a variety of methods are used to protect the eggs. Cardinalfish (Apogonidae) and jawfish (Opisthognathidae) are oral incubators. The eggs are held in the mouth of the adult until hatching.[7,44-49] Stargazers (Dactyloscopidae) are known to brood their eggs under the pectoral fin of the male.[46] Blennies (Blenniidae) and gobies (Gobiidae), among others, lay their eggs inside semienclosed structures and the males guard the eggs.[50,51] The clingfish *Gobiesox strumosus* (Order Gobiesociformes) also deposits its eggs in semi-enclosed structures with the males guarding the nests.[52,53] The puffer *Tetraodon fluviatilis* (Order Tetraodontiformes) deposits adhesive eggs on stones and the males guard the eggs until hatching.[7]

Single eggs not receiving parental care — This strategy, which is known to occur in most orders of teleost fishes, also takes a variety of forms.

The grunion *Leuresthes sardinia* (Order Atheriniformes) deposits its eggs in beach sand at high tide. The hatching embryos are swept out to sea by subsequent high tides.[54] *Fundulus* spp. (Order Atheriniformes) attach their eggs to blades of grass at high tide. These eggs may be out of the water for extended periods of time and are able to resist dessication. When the embryo has developed to the point of hatching, immersion by a subsequent high tide is apparently the cue for hatching to take place.[55,56] Some deep-water fishes in various orders appear to release their eggs at or near the bottom with eggs and/or larvae ascending through the water column to become pelagic.[6] Some fishes which are coastal dwellers as adults migrate offshore to spawn.[57-77] These fish all produce pelagic eggs. A few fish, such as the red drum *Sciaenops ocellatus* which range well offshore as adults, migrate inshore to spawn.[78] Many fish spawn in place or have only limited migrations to more advantageous places in their habitat.[68,79-81]

III. METHODS OF INTRODUCTION INTO THE PLANKTON

Although there are a variety of ways for eggs and larvae to enter the planktonic community in estuarine and marine waters, they are not as diverse as the methods of reproduction or incubation of the eggs. Nevertheless, there are enough different mechanisms used to insure the survival of the eggs and larvae to make a brief discussion worthwhile.

Some fishes of the order Atheriniformes, which lay their eggs on grass blades or vertical surfaces, produce eggs which are able to withstand long periods of dessication and which can delay hatching for extended periods of time. This ability to delay hatching probably is an adaptation which allows the embryos to hatch only when they are once again submerged, thereby insuring survival of the young.[55,82] Little is known of the behavioral mechanisms associated with hatching in the embryos of oviparous fishes which are protected by the parents. However, studies on the perciform fish *Blennius pholis* indicate that hatching is greatest during daylight hours and that the young are phototrophic moving to the surface waters shortly after hatching where they begin feeding on larval invertebrates. Since larval *B. pholis* feed primarily during the day, and since the larvae have little or no yolk reserve at hatching, this behavior probably insures their survival.[83] The ability to synchronize hatching with time of day may also be used by species which seem to hatch primarily at night.[84,85] Some species such as the damsel fish *Abudefduf zonatus* seem to synchronize hatching with the result that more larvae are hatched than can be eaten by predators on the reef before the larvae are dispersed by currents.[86] Larvae of the grunion *Leuresthes sardinia* which incubate in beach sands above the low tide mark, apparently hatch on a high tide and are swept out to sea by wave and current action.[54] The gasterosteiform fish *Aulorhynchus flavidus*, which lays its eggs in kelp beds with the male guarding the nest, generally has larvae which are only briefly planktonic. However, these fish apparently lay their eggs in portions of the kelp beds which are subject to storm disruption so that their eggs commonly break free under heavy wave action and become part of the planktonic community. This may be a mechanism which insures dispersal of this species across channels and open sand areas where they are not normally found.[43] At least some deepwater fish produce eggs and/or larvae which ascend through the water column until they are in at least the upper 200 meters of the water column where available food is more abundant.[6,87] Many fish which are found in freshwater (e.g., the eels *Anguilla* spp.) or in coastal waters (e.g., the mullets *Mugil* spp.) as adults, migrate well offshore to spawn. This migration may be to reduce predatory activity on the eggs and larvae,[81] may be to insure dispersal of the species,[80] or, in the case of fish with leptocephalus larvae (all eels, the tarpon *Megalops atlanticus*, the lady fish *Elops* spp. and the bone fish *Albula vulpes*), it may be because the larvae are nearly isosmotic with their environment and require the higher salinities and more stable environment of the open ocean.[88,89] Some of these fish migrate only a few miles offshore to spawn, but the European eel *Anguilla anguilla* migrates from the freshwaters of Europe to an area between Bermuda and the Bahamas in the western Atlantic to spawn and their larvae take as much as three years to make the return trip to Europe. Some species in the families Sciaenidae, Sparidae, Engraulidae, Gerridae and Pomadasyidae migrate inshore to spawn using the rising tides to transport their eggs and larvae into the estuaries which are the nursery grounds for the young.[78,80] Reef fish, in particular coral reef fish which do not migrate in order to spawn, generally do not merely release their eggs wherever they happen to be. Rather, they tend to move to areas of the reef which are near a break in the reef through which there is substantial current flow or which is downcurrent and seaward of the reef. These locations do not seem to be chosen arbitrarily since the same location is used over a period of years. Spawning is synchronized with current flow through the break in the reef or with seaward flow during tidal changes. The males and females generally swim rapidly toward the surface releasing their eggs and sperm at the peak of the rush. This activity gets the fertilized eggs up into the water column where they will be swept off the reef and away from the heavy concentration of predators on the reef.[68,79-81,90] Many fish which are mass spawners seem to have a propensity for spawning at dusk or at night. This propensity may reduce predation on the eggs when they are still highly concentrated,[80] and it may reduce predation on the adults which have schooled for spawning and are thus more vulnerable than usual to predators.

The reasons for mass broadcast of eggs and the introduction of eggs and larvae into the plankton by coastal fishes which move offshore to spawn and by reef fishes which utilize currents to move their eggs and larvae away from the reef have been debated.[80,81] The two primary arguments are that such activities are either to disperse the eggs and larvae over wide areas, or to avoid predation. It is true that dispersal is an important result of the mass broadcast of eggs,[75,91-104] and it is also true that getting the larvae away from the areas of high predator activity is an important effect,[80,105-107] but the assumption that either effect was the primary evolutionary motive force producing these effects to the exclusion of the other or to the exclusion of some other cause, is unverifiable and is challenged by the behavior of those fishes which are also mass broadcasters of eggs but which move into, or never leave, coastal areas or reefs and which are still successful at surviving as species and which are also widely dispersed. It is sufficient to note that both of these effects are important in the life history of the fish using them.

IV. EGG AND LARVAL DISPERSAL

Planktonic eggs and early stage larvae are at the mercy of the seas, and are moved wherever water currents and wind-driven currents carry them. At first this current controlled movement may seem haphazard with the probability of larvae arriving at suitable nursery grounds being left up to chance. Indeed, random transportation by currents may increase the area of distribution of fish outside a preexisting range,[43,93,103,108-110] and may be important for the distribution of deep-sea fishes.[87,100,104,111-114] Random transportation may also result in dispersal within a preexisting range of the young where adults do not migrate or do not migrate long distances to spawn. Fishes of the families Carapidae and Branchiostegidae are widespread as adults, yet do not move great distances as adults. Their widespread occurrence as adults is the result of the planktonic stage of the young.[91,101,108] More commonly, adult fish spawn in locations where the prevailing water and wind-driven current patterns will usually insure that a large percentage of the larvae will have access to appropriate habitat when they settle out of the plankton.[98] *Spratus spratus* (family Clupeidae), an important commercial fish in Norway and Sweden, migrates to restricted areas off Sweden to spawn. The prevailing currents in the area distribute the eggs and developing larvae throughout the Skagerrack and Kattegat areas off Sweden and Norway where they enter the estuaries and coastal areas which are their nursery grounds.[92] Adult cod (*Gadus morhua*) are known to migrate to specific spawning grounds off Norway. The prevailing currents from this area distribute the eggs and larvae to their coastal nursery grounds.[112]

It might appear that stocks of fish spawning on isolated shallow banks surrounded by deep oceanic waters would be more vulnerable to mortality due to drift into unsuitable areas than would larvae of a species like *Spratus spratus* which spawns in a more confined area. In most cases this is generally not true. *Melanogrammus aeglefinus* (Family Gadidae) is an important commercial fish in the North Sea. One of the large stocks of this species is found on the Faroe Plateau between Scotland and Iceland. This area has an eddy system which maintains a planktonic population in the area of the Faroe Plateau which is quite different from the typical oceanic planktonic population of the surrounding waters. As a result of this eddy system, the planktonic stage of *M. aeglefinus* is generally maintained over the Faroe Plateau where they settle out and eventually become part of the fishery.[113] The same type of situation seems to occur in the area of the Georges Bank off the east coast of North America where a prevailing eddy system maintains the planktonic population over the Banks separate from that of the surrounding oceanic waters.[114,115] Eddies may also play an important role in maintaining fish stocks on coral reefs.[106]

Many species of coastal fish which move offshore to spawn also depend on currents to return their larvae to the nursery grounds inshore or on the shallow shelf. The European eel *Anguilla anguilla* and the American eel *A. rostrata* migrate to an area known as the Saragasso Sea which lies between Bermuda and the Bahamas in order to spawn. Both *A. anguilla* and *A. rostrata* utilize the Gulf Stream as a distribution mechanism. Exactly how the two species are returned to their native areas without significant mixing of the stocks is unknown, but the separation may be a product of boundary layers within the Gulf Stream. *A. anguilla* is certainly the champion migrator as it moves across 90° of longitude to return from the western Atlantic to the estuaries of Europe.[116] Other species which move far offshore to spawn (e.g., the marine and estuarine eel *Myrophis punctatus* and the ladyfish *Elops saurus*) utilize long-distance current transport to maintain their stocks throughout extensive areas, and may also rely on currents to maintain separate stocks in areas such as the Gulf of Mexico. Even deep-sea fishes may have their distributions limited by prevailing currents and boundary layers.[100,104]

Many fish which use coastal areas as a nursery ground, or which live their entire life (except for spawning) in the coastal areas, move offshore to spawn, but still remain in an area where the planktonic community is different from that of the open ocean.[97] The spawning grounds for these fish tend to be seasonally and spatially determined by the prevailing water and wind-driven currents which will return the young to their nursery grounds,[97,116,117] or which will place the young in a position to move inshore using onshore currents, Eckman drift or by active swimming.[60-64,95,116-124] Bluefish (*Pomatomus saltatrix*) spawn offshore. The eggs and larvae are transported by coastal currents which maintain a salinity and temperature regime conducive to the young bluefish. When the bluefish larvae settle out of the plankton, they move inshore to the estuarine nursery ground.[121,125] *Brevortia tyrannus* (Family Clupeidae) larvae drift passively in the offshore currents until they are ready to enter the estuary which they reach using onshore drift.[122] The dover sole (*Microstomus pacificus*) which may spend as much as a year in the plankton as larvae, also apparently uses onshore currents to reach their nursery grounds.[123] Year class strength of the Pacific hake (*Merluccius productus*) is high in years when onshore Ekman drift is strong, and low in years when Ekman drift is primarily offshore.[117]

The use of onshore currents to reach a nursery ground is not always fortuitous. At least some fish appear to be thermotaxic, using the temperature gradients between inshore and offshore waters as a guide for movement. The sea chub *Girella nigricans* is thought to actively seek temperature gradients as a cue for onshore movement. When upwellings disrupt the temperature gradients, year class strength is reduced.[119] Whether this is a common method of determining which currents to use to reach shore is unknown, but many species of fish larvae and juveniles are known to move up and down in the water column to find the onshore currents.[98,119,121,124,126,127] Even fish which spawn near or in the mouth of estuaries may rely on tidal stream transport as a means of entering their nursery grounds deeper in the estuary.[78,120,127]

Utilization of onshore currents to reach coastal areas, and tidal currents in coastal areas, not only provides a directional cue for reaching a nursery ground, but also provides a significant savings in energy, especially for smaller fish which can greatly increase their range when making use of these currents.[127]

The planktonic stages of coral reef fish which are swept from the reef by currents and which may stay at sea for several months,[105] are generally also returned to the reef by currents when they are ready to settle out of the plankton. Some larvae may be maintained in eddy systems around the reefs where they were spawned so that they return to their parental reef when they settle out of the plankton;[106] other larvae may be carried by currents so that they colonize coral reefs downstream from their parental reefs.[102]

While utilization of prevailing currents generally insures that a sufficiently high percentage of larvae are able to reach their native habitat to insure an adequate stock for spawning in future years, the system is far from perfect. Many larvae are lost to the system by being transported entirely out of their adult range,[103,109,128] or out of the range of their nursery grounds.[97,113,114,117,121] Many larvae may be transported to the edge of their range where they are able to colonize, but even if they spawn as adults, their larvae may be lost to the system through transport out of the area into waters where they cannot survive as adults. This transport into peripheral areas of their range accounts for some of the tropical fauna found in the northern Gulf of Mexico which has been transported into the Gulf of Mexico from the Caribbean by the loop current,[96,99,129,130] and also accounts for some of the tropical fauna found on the southeastern coast of Florida,[131] and for some of the tropical fauna in the Bahamas and Bermuda. Transport to peripheral areas is also responsible for the presence of a pelagic tropical fauna in the Australian Bight.[132] Even in areas where the normal current patterns generally insure a high percentage of larval recruitment into the nursery ground, a large proportion of the larval population may be lost to the system during years when winds of unusual strength or duration interrupt the normal current paths.[113,117,122,126,133,134]

V. LIMITING FACTORS AFFECTING SURVIVAL OF LARVAL FISHES

Even if current conditions are generally favorable for the dispersal of eggs and larval fish, their survival is not assured. There are still other factors which affect larval survival. A few of these will be discussed briefly. References provided in the discussion will allow the interested reader to undertake a detailed study of the subject.

A. Effects of Human Activity

Little is known about the direct effects of man made pollution on planktonic fish eggs and larvae; however, some limited information is available. Low oxygen levels, whether they occur naturally or are caused by man, result in embryonic mortality due to the lack of respiratory pigment in fish eggs,[135] or they may cause abnormal embryogenesis.[136] Hydrogen sulfide, which can enter the water column in high concentrations as a result of dredging activities, is known to significantly reduce hatching success in freshwater fish,[137] and can reasonably be presumed to do the same in marine and estuarine fish. Some heavy metals such as cadmium which can enter marine and estuarine waters at levels higher than normally occurring background levels as a result of industrial activity have been shown to result in embryonic mortality in the herring *Clupea harengus* at levels of 5ppm, and to cause substantial reduction of viable hatch at concentration levels higher than 0.1ppm.[138] The eggs of the serranid *Dicentrarchus labrax* have been shown to be sensitive to copper at 10ppm, a level sometimes reached in coastal pollution.[139] The eggs of the cod *Gadus morhua* are sensitive to copper and the anionic surfactant LAS at 0.01ppm and to zinc at 1ppm, all of which are common pollutants.[140] Cytogenetic damage to developing embryos of the Atlantic mackerel *Scomber scombrus* have been correlated with the presence of various heavy metals and toxic hydrocarbons in the waters of the New York Bight.[141] Soluble components of oil have been shown to be toxic to fish eggs and larvae.[142-144] Under the right conditions, even a small oil spill could adversely affect year class size of a fish species. A major oil spill in one of the current systems such as an eddy over the Faroe Plateau in the North Sea could be devastating to a year class.

B. Effects of Temperature and Salinity

Most marine and estuarine fish eggs can tolerate a fairly wide range of salinities during development, although fertilization rates and hatching success may be significantly lowered in salinities below those normally encountered by the species. Newly hatched larvae, however, may be very sensitive to salinity variations, even in those species which normally encounter wide variations in salinity as adults. Temperature is generally much more significant for embryonic development, hatching success, and larval development that is salinity. Eggs and larvae of the Pacific sardine *Sardinops sagax* do not survive below 13°C,[145] a temperature sometimes reached in the California Current areas and which may result in delayed spawning. The petrale sole *Eopsetta jordani* is known to have an optimum developmental range of 6 to 7°C in salinities of 27 to 31 ppt with only minimal hatching success outside the ranges of 5 to 8°C and 20 to 35 ppt.[146] In a study of the occurrence of bluefish (*Pomatomus saltatrix*) eggs and larvae, it was discovered that eggs were never found in water with a salinity lower than 26.6 ppt and a temperature lower than 17°C. A maximum number of *P. saltatrix* eggs and larvae were taken in water with an average salinity of 31.1 ppt and an average temperature of 26.8°C. That this correlation was probably not a spurious one was supported by the very low number of eggs taken during a month of unseasonably cool weather during the normal spawning season.[121] The belonid *Belone belone* is known to hatch in a salinity range of 10 to 45 ppt and a temperature range of 12 to 24°C with an optimum range of 34 to 37 ppt salinity and 17 to 19°C. While salinity levels were shown to play a role in survival of *B. belone* eggs and larvae, temperature was the more important of the two factors.[147] Spawning may begin at temperatures of 6 to 8°C for *Clupeonella delicatula*, but survival is very low. Optimum survival occurs when temperatures are 15 to 18°C and salinity is below 9 ppt. At salinities above 10 ppt survival drops off sharply.[135] The Pacific cod *Gadus macrocephalus* has an optimum temperature range for embryonic development and hatching success of 3 to 5°C. Within this optimal temperature range, salinity and oxygen levels over a wide range of conditions seemed to have little effect on developing embryos. Larvae hatched in the optimal temperature range were not only more numerous than those hatched outside the temperature range, but were larger as well.[146] *Solea solea*, the Dover sole, will hatch in temperatures of 7 to 16°C, but first feeding has not been observed below 12°C.[148] Recruitment models have shown a positive correlation between sea surface temperature and recruitment success, up to an optimal sea surface temperature, for Pacific mackerel *Scomber japonicus*.[149] Laboratory studies of the red drum *Sciaenops ocellatus* produced successful hatching in salinities of 10 to 40 ppt and temperatures of 20 to 30°C, but found that optimum salinities and temperatures were 30 ppm and 25°C. It was concluded that temperature was the more important of the two factors, and that unseasonably cool temperatures during the spawning season could result in a poor year class.[150]

In contrast to most marine fish larvae, the leptocephalus larval form which characterizes the orders of the elopocephala are tied to the open seas with their higher, more stable salinities. The weakly developed integument and limited osmoregulatory ability of leptocephali confine them to the open ocean until they are near metamorphosis.[88,89] Little is known about the effects of temperature on eggs and larvae of the elopocephalans, but it is suggestive that the majority of the species are found in tropical and subtropical waters.

Salinity concentrations outside of the optimal range for fish larvae require increasing utilization of energy for osmoregulation. The energy used for osmoregulation would, under better circumstances, be used for prey capture, predator avoidance, and growth. As a consequence, growth is slowed, survival is reduced and, if the limits of osmoregulatory ability are reached, the organism dies. Temperatures below the optimum levels reduce activity. At some point, the animal is no longer able to obtain enough food to sustain life, or life functions cease altogether as a direct result of the temperature.[146,150] At higher than optimum temperatures, greater food intake is required in order to meet energy requirements. Larvae in this situation are more vulnerable to food density dependent mortality.[146]

C. Effects of Predation

Little has been written about the effects on year class size of predation on fish eggs and larvae. The high rates of mortality of eggs and early larval stages (as high as 95% per day, but more typically 30 to 40% per day[151]) suggest that predation must be intense since the eggs and early larvae utilize yolk reserves as an energy source and do not feed and thus are not subject to food density dependent mortality. Pelagic invertebrate predators such as ctenophores, chaetognaths, medusae, siphomedusae, euphausids, copepods, amphipods, and dinoflagellates have all been reported to consume fish eggs and early larvae.[136,151-156] Ctenophores filter large volumes of water and might have a deleterious effect on a year class if newly spawned eggs which were still highly concentrated should encounter a large group of ctenophores, but generally ctenophores are considered competitors for food resources with larval fish rather than predators on eggs and early stage larvae.[157] Chaetognaths are known to feed on fish eggs and early larvae, but they are thought to be responsible for only a small percentage of daily larval mortality.[151] Information obtained from laboratory studies indicate that euphausids and calanoid copepods are potentially serious predators of fish eggs and larvae;[158,159] but no food studies in the field demonstrate whether this potential predation actually occurs at high levels. Some hyperid amphipods have been implicated as serious predators on fish eggs and early larvae,[153,154] but the impact of their predation on a year class has not been assessed. Only the dinoflagellate *Noctiluca*, which is known to prey on the eggs of anchovies, has been implicated as possibly affecting a year class strength by its predation.[152]

Undoubtedly these invertebrate predators, separately and in concert, are responsible for a significant portion of daily mortality of fish eggs and early larvae. If a large patch of invertebrate predators encounter a patch of fish eggs or early larvae, that patch of eggs and larvae might well be eliminated; but currently there is no data which lends strong support to the idea that predation by invertebrates is a significant factor affecting year class strength of a species. Results of a few recent studies suggest that fish may have evolved a mechanism which minimizes the potential impact of zooplankton predators. A study of the co-occurrence of *Engraulis mordax* eggs and larvae with invertebrate predators showed an inverse correlation between patches of predators and patches of anchovy larvae. If this correlation is not a spurious one, it may be that fish isolate their eggs and larvae from invertebrate predators by spawning in areas where patches of potential invertebrate predators are absent.[155,156] In any event, the size and speed of most invertebrate predators limits their potential impact on year class size to eggs and early larvae.[151]

The predators having the greatest impact on fish eggs and larvae are probably schooling juvenile and adult fish. Gut content analysis of schooling pelagic juvenile and adult fish indicate the presence of fish eggs and larvae in many species, at least occasionally, but it is difficult to determine the exact level of their predation on fish larvae because larvae decompose rapidly in the gut and may be seen only as unrecognizable animal matter.[160] Although precise data on the impact of predation on fish eggs and larvae by schooling pelagic fish are lacking, the data that is available is suggestive. Studies on gravid *Menidia peninsulae* indicate that larval silversides are an important part of their diet.[161] Larval fish, including a high percentage of the larvae of their own species, are known to make up as much as 83% of the diet in young mackerel (*Scomber scombrus*) in the North Atlantic.[162,163] Numerous reef species are known to prey heavily on the eggs and larvae of fish.[90,164,165] Much of this predation is on newly spawned eggs as they are released and still highly concentrated.[90,164] While these predators are general filter feeders, they seem to be attracted to spawning sites specifically for the purpose of feeding.[164] In at least some species of fish, it has been hypothesized that cannibalism by the juveniles of a species on the eggs and larvae of its own species is extensive enough to account for most of the egg and larval mortality of that species.[163,166] The northern anchovy *Engraulis mordax* is known to prey on its own eggs and may account for as much as 32% of the daily mortality of the species.[160] The clupeiform fishes are probably the most important predators of fish eggs and larvae because they are abundant, planktivorous fish which travel in large schools,[151,160] and they are known to be capable of having a significant impact on zooplankton community structure.[151,167]

Planktivorous fish are known to be important predators of fish eggs and larvae, and a large school may well eliminate a patch of fish eggs or larvae, but as with invertebrate predators, there is currently no data which indicate that year class strength of a species is adversely affected by this predation.

D. Food and Feeding Effects

It may seem to be a truism to state that availability of an adequate food supply is a limiting factor since it is obvious that if fish don't have food, they can't survive. It might seem enough to say that in temperate zones larvae hatch during seasons when production of phytoplankton and zooplankton is particularly high. However, prey availability involves more than just spawning in seasons of high primary productivity. It is necessary that prey of the right sort at the right sizes be in the right concentrations in the right places if larval fish are to grow and survive. The fact that all these conditions are not always met, even during seasons and in general areas of high primary productivity, results in prey availability being a major factor affecting year class strength of a species.[116,149,151,168-179]

Even if a patch of fish larvae encounter a large patch of potential prey organisms it is no guarantee that larval survival will be high. It is necessary that the prey organism be a species which will provide adequate nutrition for the larvae since all the potential prey items which can be eaten by fish larvae are not nutritionally adequate for larval growth and survival. In 1975 during the spawning season of the northern anchovy *Engraulis mordax* a concentrated bloom of the thecate dinoflagellate *Gonyaulax polyedra* was the major food item in the region of larval anchovy occurrence.[173,178,179] The fact that the 1975 year class of anchovies was one of the worst on record despite the presence of many potential prey items can be directly attributed to the fact that *G. polyedra* is a very poor nutritional source for *E. mordax* larvae.[179] Some prey items which are adequate nutritional sources for one stage of larval growth may not be adequate for a later growth stage.[151] A species which is an adequate nutritional source for larval fish may initially be present in a spawning area, but events such as upwellings may result in the nutritionally adequate species being replaced by nutritionally inadequate species such as diatoms even though primary productivity may remain high or may increase.[179]

Available prey organisms must be of the right size for the feeding stage of the larvae. Prey organisms which are too large for the mouths of the particular larval feeding stage are obviously of no value to the larvae no matter what their nutritional content may be. Organisms below an optimal size may also be inadequate as prey items either because they cannot be seen, or because the larval fish cannot catch enough of them to get adequate nutrition even when the prey organisms are highly concentrated.[151] Laboratory tests have shown that the optimum body width of prey for first feeding larvae is 50 to 200 μm.[151,169,175] Copepod naupliis are extremely abundant in marine and estuarine waters, and their size range overlaps this optimal prey size range.[171] Copepod nauplii are also known to be excellent nutritional sources for fish larvae. It is not surprising then that they are one of the most common items found in the guts of healthy larvae. Species of dinoflagellates, which are good nutritional sources for first feeding larvae when they are larger than 50 μm, are inadequate to support larval growth when they are 40 μm because the energy required to catch and consume them exceeds the energy the larvae are able to obtain from them.[151,169] It is known that the volume of water searched by first feeding fish larvae is very small. Laboratory studies have shown that first feeding fish larvae offered prey items within the optimal size range do not strike at prey items which are farther than one body length away from the fish and that the striking distance may be as little as 0.2 of the fish larva's body length.[151] A correlation between prey size and larval fish striking distance has not yet been tested for, but it is probable that striking range is determined, at least in part, by the size of the prey item. Smaller items, even though nutritionally adequate, are harder to see.[151]

Laboratory studies have shown that first feeding fish larvae must consume enough prey to have a growth rate of 15 to 20% per day in order to survive.[175] Since first feeding fish larvae only strike at prey items which are very close to them, it is apparent that prey densities must be high enough to insure frequent encounters with prey. Laboratory studies and computer models confirm that there are critical prey densities which are necessary for good larval fish growth and survival.[149,170,172,174,175] Initial laboratory studies indicated that densities of 1,000 to 4,000 microcopepods/ℓ were necessary for good larval survival, a level not routinely reached in oceanic waters even in most zooplankton patches. More recent, carefully controlled laboratory studies have shown that stocking densities similar to those found in zooplankton patches are sufficient for good larval growth and survival.[174,175] It is particularly important that high prey densities be present when fish larvae are ready to start feeding. Only a small percentage of fish larvae are successful at first feeding when prey densities are at levels similar to those found outside of zooplankton patches (10 to 1000 microcopepods/ℓ), and that larvae which do not begin feeding at these low levels cannot survive more than 5 to 6 days. Of those larvae which do begin feeding at low prey levels, only a small percentage are able to obtain enough food to meet their nutritional needs. Consequently, even most successful first feeders at low prey densities ultimately starve. When prey densities are above the critical minimum, there is a much higher percentage of first feeding success and, consequently, a much higher survival rate.[175]

Food densities necessary for larval survival may routinely occur in enclosed areas such as bays and lagoons,[151] but the average density which occurs in the open ocean is far below the optimum densities. Fortunately for larval fish, zooplankton in the open sea are not evenly distributed, but occur in patches where the density is high enough to support good larval growth and survival.[151,172,175] When larvae encounter a patch of food, they are able to maintain themselves within that patch.[172] Survival of larvae then is related not only to high zooplankton productivity, but also to the spatial occurrence of zooplankton patches. A study of northern anchovy (*Engraulis mordax*) larvae showed that as many as 60% of the larvae in one patch were starving, yet all the larvae in tows a few miles away were in excellent condition.[176] The results of this study indicate that as much as 40% of daily larval mortality could be due to starvation. On a much higher scale, the fishery for Japanese sardines (*Sardinops melanosticta*) off the coast of Japan was severely depressed for several years due to a cell of cold water which remained adjacent to the coast. When this cold water cell broke down, the Kuroshio current moved in closer to the coast bringing warm, rich waters which resulted in a sharp increase in the number of copepod nauplii with an attendant sharp increase in the size of the sardine population available to the fishery.[177]

VI. CONCLUSIONS

Estuarine and marine fish have evolved a variety of mechanisms for insuring that a sufficient number of their young survive to grow to adulthood to replenish the species stock. Some species produce a small number of young but insure that, on average, a high percentage of their young will survive. These young are provided with a nutritional source, either from maternal tissues or from a large quantity of yolk in a large egg, which allows the young to develop to a larger size before hatching. The larger size young are less vulnerable to predation, are capable of taking a large variety of prey items immediately after hatching, and are capable of greater physiological control. In some of these species, the young are also protected in some manner by one or both of the adults. Most of these species either do not have a planktonic larval stage or have a short residence time in the plankton. While these modes of reproduction insure high survival rates, the short planktonic residence time of their larvae generally reduces dispersal range for the species. Many species produce a large number of eggs which are unprotected and which have only a small yolk reserve. A very high percentage of these fertilized eggs never develop to sexual maturity, but they are readily dispersed over a large area and are the first species to colonize a new area or an area with low population densities.

The planktonic egg and/or larval stage is inextricably linked to the survival of most species of fish. It removes them from areas of intense predation such as the coastal areas, coral reefs and benthic habitats of their adult stage; reduces competition for available food resources by spreading the young out over much wider areas than would be possible if all species spawned in place; insures that localized biological perturbations do not wipe out an entire year class either through direct effect on the species or through effects on its food source; provides an area of high food productivity for larvae such as the larvae of deep dwelling fish which live in areas of low productivity as adults; provides a means of return to nursery grounds for species whose adults migrate to spawning grounds, or a means of widespread dispersal which is particularly important for benthic dwelling species which do not migrate long distances as adults; and, in some cases, provides the eggs and larvae with a physiologically suitable medium which is more hospitable than the one in which they may reside as adults.

While the planktonic stage is necessary for the dispersal and survival of many species, it is not without its dangers. Fish eggs and larvae, particularly in coastal areas, are subject to the hostile conditions caused by man's pollutants. This source of mortality has not been shown to be particularly significant in the open ocean, although the potential is certainly there. Naturally occurring environmental conditions, such as unseasonally cold or warm years, unusually wet, or, in coastal areas, dry years, winds which are unseasonally strong or which are of longer than normal duration, and which consequently change current patterns normally used for distribution of eggs and larvae; and upwellings which alter current patterns or which affect food sources, are all major causes of egg and larval mortality, and are known to account for bad year classes for some species. Predation by both vertebrate and invertebrate species is known to be responsible for a large percentage of the daily mortality of fish eggs and larvae, but there is little evidence to show that predation is responsible for poor year classes. The major cause of larval mortality is starvation. In order for larvae to have good growth and survival rates, the right sort of food at the right sizes must be in the right concentrations in the right places. The same environmental conditions which directly affect eggs and larval fish can prevent an adequate food supply from being present by limiting productivity, by dispersing concentrated patches of zooplankton, or by replacing an adequate food source with an inadequate one.

Different species of fish eggs and larvae are differently adapted to their environment and are thus more vulnerable to some causes of mortality than others. Many of the factors limiting good growth and survival act synergistically. Many factors affecting growth and survival are certainly, as yet, unknown. Much remains to be learned about how, where and when larval fish grow, how fish stocks are regulated, and how man affects survival of the various fish species.

REFERENCES

1. Bennett, D. E. and Wydoski, R. S., Biology of the red tail surfperch (*Amphistichus rhodoterus*) from the central Oregon coast, *U.S. Fish Wildl. Serv. Tech. Pap.*, 90, 1, 1977.
2. Hildebrand, S. F., Family Elopidae, in *Fishes of the Western North Atlantic*, Olsen, Y. H., Ed., Mem. Sears Found. Mar. Res. 1, Yale University, New Haven, Conn., 1963, III.
3. Amoroso, E. C., Viviparity in fishes, *Symp. Zool. Soc. Lond.*, 1, 153, 1960.
4. Hoar, W. S., Viviparity and gestation, in *Fish Physiology III*, Hoar, W. S. and Randall, D. S., Eds., Academic Press, New York, 1969, 20.
5. Cohen, D. M., A review of the ophidioid fish genus *Oligopus* with the description of a new species from West Africa, *Proc. U.S. Natl. Mus.*, 116, 1, 1964.
6. Mead, G. W., Bertelsen, E., and Cohen, D. H., Reproduction among deep-sea fishes, *Deep-Sea Res.*, 11, 569, 1964.
7. Breeder, C. M. and Rosen, D. E., *Modes of reproduction in fishes*, Natural History Press, Garden City, N.Y., 1966, 941.
8. Cohen, D. M. and Nielsen, J. C., Guide to the identification of the fish order Ophidiiformes with a tentative classification of the order, *NOAA Tech. Rept., NMFS Circ.*, 471, 1978.
9. Gilbert, P. W. and Schlernitzauer, E. A., The placenta and gravid uterus of *Carcharhinus falciformis*, *Copeia*, 451, 1966.
10. Bigelow, H. B. and Schroeder, W. G., Sharks, in *Fishes of the Western North Atlantic*, Parr, A. E. and Olsen, Y. H., Eds., Mem. Sears Found. Mar. Res. 1, Yale University, New Haven, Conn., 1948, 59.
11. Stribling, M. D., Hamlett, W. C., and Wourms, J. P., Developmental efficiency of oophagy, a method of viviparous embryonic nutrition displayed by the sand tiger shark (*Eugomophodus taurus*), *Bull. S.C. Acad. Sci.*, 42, 111, 1980.
12. Gilbert, P. W., Patterns of shark reproduction, *Oceanus*, 24, 30, 1982.
13. Mendoza, G., Structural and vascular changes accompanying the respiration of the proctodeal processes after birth in the embryos of the Goodeidae, a family of viviparous fishes, *J. Morphol.*, 61, 95, 1937.
14. Mendoza, G., The reproductive cycle of the viviparous teleost *Neotoca bilineata*, a member of the family Goodeidae, *Biol. Bull.*, 76, 359, 1939.
15. Mendoza, G., Adaptations during gestation in the viviparous cyprinodont teleost *Hubbsina turneri*, *J. Morphol.*, 99, 73, 1956.
16. Mendoza, G., The fin fold of *Goodea luipoldii*, a viviparous cyprinodont teleost, *J. Morphol.*, 103, 539, 1958.

17. Mendoza, G., The ovary and anal process of *"Characodon" eiseni*, a viviparous cyprinodont teleost from Mexico, *Biol. Bull.*, 120, 303, 1965.

18. Turner, C. L., Viviparity superimposed upon ovo-viviparity in the family Goodeidae, a family of cyprinodont fishes of the Mexican Plateau, *J. Morphol.*, 55, 207, 1933.

19. Turner, C. L., The absorptive processes in the embryos of *Parabrotula dentients*, a viviparous deep-sea brotulid fish, *J. Morphol.*, 59, 313, 1936.

20. Turner, C. L., The trophotaeniae of the Goodeidae, a family of viviparous cyprinodont fishes, *J. Morphol.*, 61, 495, 1937.

21. Turner, C. L., Adaptations for viviparity in embryos and ovary of *Anableps anableps*, *J. Morphol.*, 62, 323, 1938.

22. Turner, C. L., The pseudoamnion, pseudochorion, pseudoplacenta and other foetal structures in viviparous cyprinodont fishes, *Science*, 90, 42, 1939.

23. Turner, C. L., Pseudoamnion, pseudochorion and follicular pseudoplacenta in poecillid fishes, *J. Morphol.*, 67, 59, 1940.

24. Turner, C. L., Follicular pseudoplacenta and gut modifications in anablepid fishes, *J. Morphol.*, 67, 91, 1940.

25. Turner, C. L., Pericardial sac, trophotaeniae and alimentary tract in embryos of goodeid fishes, *J. Morphol.*, 67, 271, 1940.

26. Turner, C. L., Adaptations for viviparity in jenynsiid fishes, *J. Morphol.*, 67, 291, 1940.

27. Turner, C. L., An accessory respiratory device in embryos of the embiotocid fish, *Cymatogaster aggregata* during gestation, *Copeia*, 1952, 146, 1952.

28. Webb, P. W. and Brett, J. R., Respiratory adaptations of prenatal young in the ovary of two species of viviparous sea perch, *Rhacochilus vacca* and *Embiotoca lateralis*, *J. Fish. Res. Board Can.*, 29, 1525, 1972.

29. Webb, P. W. and Brett, J. R., Oxygen consumption of embryos and parents, and oxygen transfer characteristics within the ovary of two species of viviparous sea perch, *Rhacochilus vacca* and *Embiotoca lateralis*, *J. Fish. Res. Board Can.*, 29, 1543, 1972.

30. Kristoffersson, R., Broberg, S., and Pekkarinen, M., Histology and physiology of embryotrophe formation, embryonic nutrition and growth in the eel-pout, *Zoarces viviparus* (L.), *Ann. Zool. Fenn.*, 10, 467, 1973.

31. Ahlstrom, E. H., Vertical distribution of pelagic fish eggs and larvae off California and Baja California, *Fish. Bull.*, 60, 107, 1959.

32. Wourms, J. P. and Bayne, O., Development of the viviparous brotulid fish, *Dinematichthys ilucoeteoides*, *Copeia*, 1973, 32, 1973.

33. Moser, H. G., Reproduction and development of *Sebastodes paucispinis* and comparison with other rockfishes off Southern California, *Copeia*, 1967, 773, 1967.

34. Atz, J. W., *Latimeria* babies are born, not hatched, *Underwater Nat.*, 9, 4, 1976.

35. Cox, P., Histories of new food fishes, II. the lumpfish, *Bull. Biol. Board Can.*, 2, 1, 1920.

36. Cox, P. and Anderson, M., A study of lumpfish (*Cyclopterus lumpus* L.), *Contrib. Can. Biol.*, 1, 1, 1924.

37. Shiogaki, M., Life history of the stichaeid fish *Opisthocentrus ocellatus*, *Jpn. J. Ichthyol.*, 29, 77, 1982.

38. Lobel, P. S. and Johannes, R. E., Nesting, eggs and larvae of trigger-fishes, *Environ. Biol. Fishes*, 5, 251, 1980.

39. Bigelow, H. B. and Schroeder, W. C., Fishes of the Gulf of Maine, *Fish. Bull.*, 53, 1, 1953.

40. Pietsch, T. W. and Grobecker, D. B., Parental care as an alternative reproductive mode in an antennariid anglerfish, *Copeia*, 1980, 551, 1980.

41. Orton, G. L., Early developmental stages of the California scorpionfish *Scorpaena guttata*, *Coepia*, 1955, 210, 1955.

42. Limbaugh, C., Life history and ecological notes on the tubenose, *Aulorhynchus flavidus*. a hemibranch fish of western North America, *Copeia*, 1962, 549, 1962.

43. Marliave, J. B., A theory of storm induced drift dispersal of the gasterosteid fish *Aulorhynchus flavidus*, *Copeia*, 1976, 794, 1976.

44. Hale, H. M., Evidence of the habit of oral gestation in the south Australian marine fish (*Apogon conspersus Klunzinger*), *S. Aust. Nat.*, 24, 1, 1947.

45. Böhlke, J. E. and Chaplin, C. C. G., Oral incubation in Bahaman jawfishes *Opistognathus whitehursti* and *O. maxillosus*, *Science*, 125, 353, 1957.

46. Böhlke, J. E. and Chaplin, C. C. G., *Fishes of the Bahamas and adjacent tropical waters*, Livingston Publishing, Wynnewood, Pa., 1968, 771.

47. Colin, P. L., Daily activity patterns and effects of environmental conditions on the behavior of the yellowhead jawfish, *Opistognathus aurifrons*, with notes on its ecology, *Zoologica*, 57, 137, 1972.

48. Charney, P., Oral brooding in the cardinalfishes, *Phaeoptyx conklini* and *Apogon maculatus* from the Bahamas, *Copeia,* 1976, 198, 1976.

49. Colin, P. L. and Arneson, D. W., Aspects of the natural history of swordtail jawfish, *Lonchopisthus micrognathus* (Poey) (Pisces: Opistognathidae), in south-western Puerto Rico, *J. Nat. Hist.,* 12, 689, 1978.

50. Peters, K. M., Reproductive biology and developmental osteology of the Florida blenny, *Chasmodes saburrae* (Perciformes: Blenniidae), *Northeast Gulf Sci.,* 4, 79, 1981.

51. Peters, K. M., Larval and early juvenile development of the frillfin goby, *Bathygobius soporator* (Perciformes: Gobiidae), *Northeast Gulf Sci.,* (in press).

52. Runyan, S., Early development of the clingfish *Gobiesox strumosus* Cope, *Chesapeake Sci.,* 2, 113, 1961.

53. Saksena, V. P. and Joseph, E. B., Dissolved oxygen requirements of newly-hatched larvae of the striped blenny (*Chasmodes bosquianus*), the naked goby (*Gobiosoma bosci*), and the skilletfish (*Gobiesox strumosus*), *Chesapeake Sci.,* 13, 23, 1972.

54. Reynolds, W. M. and Thompson, D. A., Responses of young gulf grunion, *Leuresthes sardinia,* to gradients of temperature, light, turbulence and oxygen, *Copeia,* 1974, 747, 1974.

55. Harrington, R. W., Delayed hatching in stranded eggs of marsh killifish, *Fundulus confluentus, Ecology,* 40, 430, 1959.

56. Taylor, M. M., DiMichele, L., and Leach, G. J., Egg stranding in the life cycle of the mummichog, *Fundulus heteroclitus, Copeia,* 1977, 397, 1977.

57. Hildebrand, S. F. and Cable, L. E., Further notes on the development and life history of some teleosts at Beaufort, N.C., *Fish. Bull.,* 24, 505, 1938.

58. Reid, G. K., Jr., An ecological study of the Gulf of Mexico fishes, in the vicinity of Cedar Key, Florida, *Bull. Mar. Sci. Gulf Caribb.,* 4, 1, 1954.

59. Caldwell, D. K., The biology and systematics of the pinfish, *Lagodon rhomboides* (Linnaeus), *Bull. Fla. State Mus. Biol. Sci.,* 2, 77, 1957.

60. Anderson, W. W., Early development, spawning, growth, and occurrence of the silver mullet (*Mugil curema*) along the South Atlantic coast of the United States, *Fish. Bull.,* 119, 397, 1957.

61. Anderson, W. W., Larval forms of the fresh-water mullet (*Agonostomus monticola*) from the open ocean off the Bahamas and South Atlantic coast of the United States, *Fish. Bull.,* 120, 415, 1957.

62. Anderson, W. W., Larval development, growth and spawning of the striped mullet (*Mugil cephalus*) along the South Atlantic coast of the United States, *Fish. Bull.,* 58, 501, 1958.

63. Arnold, E. L., Jr. and Thompson, J. R., Offshore spawning of the striped mullet, *Mugil cephalus,* in the Gulf of Mexico, *Copeia,* 1958, 130, 1958.

64. Caldwell, D. K. and Anderson, W. W., Offshore occurrence of larval silver mullet, *Mugil curema,* in the western Gulf of Mexico, *Copeia,* 1959, 252, 1959.

65. Springer, V. G. and Woodburn, D. K., An ecological study of the fishes of the Tampa Bay area, *Fla. Board Conserv. Mar. Res. Lab. Prof. Ser.,* 1, 1, 1960.

66. Alexander, E. C., A contribution to the life history, biology and geographical distribution of the bonefish, *Albula vulpes* (Linnaeus), *Dana Rep.,* 53, 1, 1961.

67. Wade, R. A., The biology of the tarpon, *Megalops atlanticus,* and the oxeye, *Megalops cyprinoides,* with emphasis on larval development, *Bull. Mar. Sci. Gulf Caribb.,* 12, 545, 1962.

68. Randall, J. E. and Randall, H. A., The spawning and early development of the Atlantic parrotfish, *Sparisoma rubripinne,* with notes on other scarid and labrid fishes, *Zoologica,* 48, 49, 1963.

69. Eldred, B. and Lyons, W. G., Larval ladyfish, *Elops saurus* Linnaeus, 1766 (Elopidae), in Florida and adjacent waters, *Fla. Board Conserv. Mar. Res. Lab. Leaf. Ser.,* 2, 1, 1966.

70. Eldred, B., Larval bonefish, *Albula vulpes* (Linnaeus, 1758) (Albulidae), *Fla. Board Conserv. Mar. Res. Lab. Leaf. Ser.,* 3, 1, 1967.

71. Eldred, B., Larval tarpon, *Megalops atlanticus* Valenciennes (Megalopidae), in Florida waters, *Fla. Board Conserv. Mar. Res. Lab. Leaf. Ser.,* 4, 1, 1967.

72. Hansen, D. J., Food, growth, migration, reproduction and abundance of pinfish, *Lagodon rhomboides* and Atlantic croaker, *Micropogon undulatus,* near Pensacola, Florida, *Fish. Bull.,* 68, 135, 1970.

73. Ogden, J. C. and Buckman, N. S., Movements, foraging groups, and diurnal migrations of the striped parrotfish *Scarus croicensis* Block (Scaridae), *Ecology,* 54, 589, 1973.

74. Miller, J. M., Nearshore distribution of Hawaiian marine fish larvae: effects of water quality, turbidity and currents, in *The Early Life History of Fish,* Blaxter, J. H. S., Ed., Springer-Verlag, N.Y., 1974, 217.

75. Leis, J. M. and Miller, J. M., Offshore distribution patterns of Hawaiian fish larvae, *Mar. Biol.,* 36, 359, 1976.

76. Houde, E. D., Leak, J. C., Dowd, C. E., and Berkeley, S. A., Ichthyoplankton abundance and diversity in the eastern Gulf of Mexico, *BLM Contract Rept. PB-299 839,* 546, 1979.

77. Smith, D. G., Early larvae of the tarpon, *Megalops atlanticus* Valenciennes (Pisces: Elopidae), with notes on spawning in the Gulf of Mexico and the Yucatan Channel, *Bull. Mar. Sci.,* 30, 136, 1980.

78. Jannke, T. E., Abundance of young sciaenid fishes in Everglades National Park, Florida, in relation to season and other variables, *Sea Grant Tech. Bull.,* (University of Miami), 11, 1, 1971.

79. Colin, P. L. and Clavijo, I. E., Mass spawning by the spotted goatfish, *Pseudopeneus maculatus* (Bloch) (Pisces: Mullidae), *Bull. Mar. Sci.,* 28, 780, 1978.

80. Johannes, R. E., Reproductive strategies of coastal marine fishes in the tropics, *Environ. Biol. Fishes,* 3, 65, 1978.

81. Barlow, G. W., Patterns of parental investment, dispersal and size among coral-reef fishes, *Environ. Biol. Fishes,* 6, 65, 1981.

82. Koenig, C. C. and Livingston, R. J., The embryological development of the diamond killifish (*Adinia xenica*), *Copeia* 1976, 435, 1976.

83. Qasim, S. Z., The spawning habits and embryonic development of the shanny (*Blennius pholis* L.), *Proc. Zool. Soc. Lond.,* 127, 79, 1956.

84. Moyer, J. T. and Bell, L. J., Reproductive behavior of the anemone fish *Amphiprion clarkii* at Jiyake-jima, Japan, *Jpn. J. Ichthyol.,* 23, 23, 1976.

85. Ross, R. M., Reproductive behavior of the anemonefish *Amphiprion melanopus* on Guam, *Copeia,* 1978, 103, 1978.

86. Keenleyside, M. H. A., The behavior of *Abudefduf zonatus* (Pisces: Pomacentridae) at Heron Island, Great Barrier Reef, *Anim. Behav.,* 20, 763, 1972.

87. Marshall, N. B., *Explorations in the Life of Fishes,* Harvard University Press, Cambridge, Mass., 1971, 204.

88. Hulet, W. H., Fischer, J., and Rietberg, B. J., Electrolyte composition of anguilliform leptocephali from the Straits of Florida, *Bull. Mar. Sci.,* 22, 432, 1972.

89. Hulet, W. H., Structure and functional development of the eel leptocephalus *Ariosoma balearicum* (De la Roche 1809), *Philos. Trans. R. Soc. Lond.,* (B) Biol. Sci., 282, 107, 1977.

90. Colin, P. L., Daily and summer-winter variation in mass spawning of the striped parrotfish, *Scarus croicensis, Fish. Bull.,* 76, 117, 1978.

91. Dooley, J. K., Systematics and biology of the tilefishes (Perciformes: Branchiostegidae and Malacanthidae), with description of two new genera, *NOAA Tech. Rept. NMFS Circ.,* 411, 1, 1978.

92. Lindquist, A., On fish eggs and larvae in the Skagerrak, *Sarsia,* 34, 347, 1968.

93. Scheltema, R. S., Dispersal of larvae by equatorial ocean currents and its importance to the zoogeography of shaol-water tropical species, *Nature (London),* 217, 1159, 1968.

94. Briggs, J. C., Tropical shelf zoogeography, *Proc. Calif. Acad. Sci.,* 38, 131, 1970.

95. Graham, J. J., Chenoweth, S., and Davis, C., Abundance, distribution, movements and lengths of larval herring along the western coast of the Gulf of Mexico, *Fish. Bull.,* 70, 307, 1972.

96. Duncan, D. C., Atwood, D. D., Duncan, J. R., and Froelich, P. N., Drift bottle returns from the Caribbean, *Bull. Mar. Sci.,* 27, 580, 1977.

97. Richardson, S. L. and Pearcy, W. A., Coastal and oceanic fish larvae in an area of upwelling off Yaquina Bay, Oregon, *Fish. Bull.,* 75, 125, 1977.

98. Brattstrom, H., The importance of water movements for biology and distribution of marine organisms, *Sarsia,* 34, 9, 1968.

99. Metcalf, W. G., Stalcup, M. C., and Atwood, D. G., Mona passage drift bottle survey, *Bull. Mar. Sci.,* 27, 586, 1977.

100. Hureau, J. C., Geistdoerfer, P., and Rannou, M., The ecology of deep-sea benthic fishes, *Sarsia,* 64, 103, 1979.

101. Olney, J. E. and Markle, D. F., Description and occurrence of vexillifer larvae of *Echiodon* (Pisces: Carapidae) in the western North Atlantic and notes on other carapid vexillifers, *Bull. Mar. Sci.,* 29, 365, 1979.

102. Grant, C. J. and Wyatt, J. R., Surface currents in the eastern Cayman and western Caribbean seas, *Bull. Mar. Sci.,* 30, 613, 1980.

103. Markle, D. F., Scott, W. B., and Kohler, A. C., New and rare records of Canadian fishes and the influence of hydrography on resident and nonresident Scotian Shelf ichthyofauna, *Can. J. Fish. Aquat. Sci.,* 37, 49, 1980.

104. Johnson, R. K., Fishes of the families Evermannellidae and Scopelarchidae: Systematics, morphology, interrelationships, and zoogeography, *Fieldiana Zool.,* 12, 1, 1982.

105. Sale, P. F., Distribution of larval Acanthuridae off Hawaii, *Copeia,* 1970, 765, 1970.

106. Emery, A. R., Eddy formation from an oceanic island: ecological effects, *Caribb. J. Sci.,* 12, 121, 1972.

107. Hamner, W. M. and Hauri, I., Surface currents of Cid Harbour, Whitsunday Island, Queensland, Australia: effect of tide and topography, *Aust. J. Mar. Freshwater Res.,* 28, 333, 1977.

108. Robins, C. R. and Nielsen, J. G., *Synderidia bothrops,* a new tropical, amphiatlantic species (Pisces, Carapidae), *Stud. Trop. Oceanogr.,* 4, 285, 1970.

109. DeSylva, D. P. and Scotton, L. N., Larvae of deep-sea fishes (Stomiatoidea) from Biscayne Bay, Florida, USA, and their ecological significance, *Mar. Biol.,* 12, 122, 1972.

110. Horn, M. H., Systematic status and aspects of the ecology of the elongate ariommid fishes (Suborder Stromateoidei) in the Atlantic, *Bull. Mar. Sci.,* 22, 537, 1972.

111. Merrett, N. R., On the identity and pelagic occurrence of larval and juvenile stages of rattail fishes (family Macrouridae) from 60°N, 20°W and 53°N, 20°W, *Deep-Sea Res.,* 25, 147, 1978.

112. Ellersten, B., Solemdal, P., Strømme, T., Sundby, S., Tilseth, T., Westgard, T., and Øiestad, V., Spawning period, transport and dispersal of eggs from the spawning area of Arcto-Norwegian cod (*Gadus morhua* L.), *Rapp. P.-V. Reun, Cons. Int. Explor. Mer.,* 178, 260, 1981.

113. Saville, A., Factors controlling dispersal of the pelagic stages of fish and their influence on survival, *Int. Comm. Northwest Atl. Fish Spec. Publ.,* 6, 335, 1965.

114. Walford, L. A., Effects of currents on distribution and survival of the eggs and larvae of the haddock, (*Melanogrammus aeglefinus*) on Georges Bank, *Bull. Bur. Fish* (U.S.), 29, 72, 1938.

115. Lough, G. G., The distribution and abundance, growth, and mortality of Georges Bank-Nantucket Shoals herring larvae during the 1975—76 winter period, *NOAA Tech. Rept. NMFS Circ.,* 415, 309, 1978.

116. Cushing, D. H., *Marine Ecology and Fisheries,* Cambridge University Press, Cambridge, 278, 1975.

117. Bailey, K. M., Larval transport and recruitment of Pacific hake *Merluccius productus, Mar. Ecol. Prog. Ser.,* 6, 1, 1981.

118. Fraser, J. H., The drift of the planktonic stages of fish in the northeast Atlantic and its possible significance to the stocks of commercial fish, Int. *Comm. Northwest Atl. Fish Spec. Publ.,* 1, 289, 1958.

119. Norris, K. S., The functions of temperature in the ecology of the percoid fish *Girella nigricans* (Ayres), *Ecol. Monogr.,* 33, 23, 1963.

120. Das, N., Spawning, distribution, survival, and growth of larval herring (*Clupea harengus* L.) in relation to hydrographic conditions in the Bay of Fundy, *Fish. Res. Board Can. Tech. Rept.,* 88, 1, 1968.

121. Norcross, W. H., Richardson, S. L., Mossman, W. H., and Joseph, E. B., Development of young bluefish (*Pomatomus saltatrix*) and distribution of eggs and young in Virginia coastal waters, *Trans. Am. Fish. Soc.,* 103, 477, 1974.

122. Nelson, W. R., Ingham, M. C., and Schaaf, W. E., Larval transport and year-class strength of Atlantic menhaden, *Brevortia tyrannus, Fish. Bull.,* 75, 23, 1977.

123. Pearcy, W. G., Hosie, M. J., and Richardson, S. L., Distribution and duration of pelagic life of larvae of Dover sole, *Microstomus pacificus:* rex sole, *Glyptocephalus zachirus;* and petrale sole, *Eopsetta jordani,* in waters off Oregon, *Fish. Bull.,* 75, 173, 1977.

124. Harden Jones, F. R., Arnold, G. P., Greer Walker, M., and Scholes, P., Selective tidal stream transport and the migration of plaice (*Pleuronectes platessa* L.) in the southern North Sea, *J. Cons. Perm. Int. Explor. Mer.,* 38, 331, 1979.

125. Kendall, A. W., Jr. and Walford, L. A., Sources and distribution of bluefish, *Pomatomus saltatrix,* larvae and juveniles off the east coast of the United States, *Fish. Bull.,* 77, 213, 1979.

126. Iselin, C. O'D., Coastal currents and the fisheries. Papers in marine biology and oceanography, *Deep-Sea Res., Suppl.,* 3, 474, 1955.

127. Weihs, D., Tidal stream transport as an efficient method for migration, *J. Cons. Perm. Int. Explor. Mer.,* 38, 92, 1978.

128. Zurbrigg, R. E. and Scott, W. B., Evidence of expatriate populations of the lanternfish *Myctophum punctatum* in the northwest Atlantic, *J. Fish. Res. Board Can.,* 29, 1679, 1972.

129. Habury, K., Hastings, R. W., Devries, D., and Massey, J., Tropical marine fishes from Pensacola, Florida, *Fla. Sci.,* 37, 105, 1974.

130. Hastings, R. W., The origin and seasonality of the fish fauna on a new jetty in the northeastern Gulf of Mexico, *Bull. Fla. State Mus. Biol. Sci.,* 24, 1, 1979.

131. Gilmore, R. G., Jr., Fishes of the Indian River lagoon and adjacent waters, *Bull. Fla. State Mus. Biol. Sci.,* 22, 101, 1977.

132. Garrey, J., Maxwell, H., and Creswell, G. R., Dispersal of tropical marine fauna to the great Australian Bight by the Leeuwin Current, *Aust. J. Mar. Freshwater Res.,* 32, 493, 1981.

133. Frederick, B. D., Ekman transport is a factor contributing to population variations of the Atlantic croaker, *Micropogon undulatus* (Linnaeus), in Chesapeake Bay, Ms. Thesis, American Univ., Wash, D.C., 1978.

134. Bolz, G. R., Lough, R. G., and Potter, D. C., Autumn and winter abundance and distribution of ichthyoplankton on Georges Bank and Nantucket Shoals, 1974—1976, with special emphasis on dominant species, *Rapp. P.-V. Reun Cons. Int. Explor. Mer,* 178, 168, 1981.

135. Pinus, G. H., Some factors influencing early survival and abundance of *Clupeonella* in the Sea of Azov, in *The Early Life History of Fish,* Blaxter, J. H. S., Ed., Springer-Verlag, N.Y., 1974, 81.

136. Hempel, G., Early life history of marine fish. The egg stage. *Sea Grant Publ.* (Wash. Univ.), 70, 1979.

137. Smith, L. L. and Oseid, D. M., Effect of hydrogen sulfide on development and survival of eight freshwater fish species, in *The Early Life History of Fish,* Blaxter, J. H. S., Ed., Springer-Verlag, N. Y., 1974, 384.

138. Rosenthal, H. and Sperling, K. R., Effects of cadmium on development and survival of herring eggs, in *The Early Life History of Fish,* Blaxter, J. H. S., Ed., Springer-Verlag, N. Y., 1974, 384.

139. Cosson, R. P. and Martin, J. L. M., The effects of copper on the embryonic development, larvae, alevins, and juveniles of *Dicentrarchus labrax* (L.), *Rapp. P. V. Reun. Cons. Int. Explor. Mer,* 178, 71, 1981.

140. Swedmark, M. and Granmo, A., Effects of mixtures of heavy metals and a surfactant on the development of cod *Gadus morhua* L.), *Rapp. P. -V. Cons. Int. Explor. Mer,* 178, 95, 1981.

141. Longwell, A. C. and Hughes, J. B., Cytologic, cytogenetic and embryologic state of Atlantic mackerel eggs from surface waters of the New York Bight in relation to pollution. *Rapp. P. -V. Reun. Cons. Int. Explor. Mer,* 178, 76, 1981.

142. Evans, D. R. and Rice, S. D., Effects of oil on marine ecosystems: a review for administrators and policy makers, *Fish. Bull.,* 72, 625, 1974.

143. Johnston, R., What North Sea oil might cost fisheries, *Rapp. P. -V. Reun. Cons. Int. Explor. Mer,* 177, 212, 1977.

144. Korn, S. and Rice, S., Sensitivity to, and accumulation and depuration of, aromatic petroleum components by early life stages of coho salmon (*Oncorhynchus kisutch*), *Rapp. P. -V. Reun. Const. Int. Explor. Mer,* 178, 1981.

145. Smith, P. E., Fisheries on coastal pelagic schooling fish, in Lasker, R., Ed., Marine fish larvae — morphology, ecology, and relation to fisheries, *Sea Grant Publ.,* (Washington University), 1981, 1.

146. Alderdice, D. F. and Forrester, C. R., Effects of salinity and temperature on embryonic development of the petrale sole (*Eopsetta jordani*), *J. Fish. Res. Board Can.,* 28, 727, 1971.

147. Fonds, M., Rosenthal, H., and Alderdice, D. F., Influence of temperature and salinity on embryonic development, larval growth and number of vertebrae of the garfish *Belone belone,* in *The Early Life History of Fish,* Blaxter, J. H. S., Ed., Springer-Verlag, N. Y., 1974, 449.

148. Irvin, D. N., Temperature tolerance of early developmental stages of Dover sole, *Solea solea* (L.), in *The Early Life History of Fish,* Blaxter, J. H. S., Ed., Springer-Verlag, N. Y., 1974, 449.

149. Parrish, R. H. and MacCall, A., Climatic variations and exploitation in the Pacific mackerel fishery, *Calif. Dep. Fish Game Fish. Bull.,* 167, 1, 1978.

150. Holt, J., Godbout, R., and Arnold, C. R., Effects of temperature and salinity on egg hatching and larval survival of red drum, *Sciaenops Ocellata, Fish. Bull.,* 79, 569, 1981.

151. Hunter, J. R., Feeding ecology and predation of marine fish larvae, in Marine fish larvae: morphology, ecology and relation to fisheries, Lasker, R., Ed., *Sea Grant Publ.* (Washington University), 1981, 34.

152. Hattori, S., Predatory activity of *Noctiluca* on anchovy eggs, *Bull. Tokyo Reg. Fish. Res. Lab.,* 9, 211, 1962.

153. Sheader, M. and Evans, F., Feeding and gut structure of *Parathemisto guadichaudi* (Guerin) (Amphipoda, Hyperiidea), *J. Mar. Biol. Assoc. U. K.,* 55, 641, 1975.

154. Westernhagen, H. von., Some aspects of the biology of the hyperid amphipod *Hyperoche medwarum, Helgol. Wiss. Meeresunters.,* 28, 1976.

155. Alvarino, A., The relation between the distribution of zooplankton predators and anchovy larvae, *Calif. Coop. Oceanic Fish. Invest. Rep.,* 21, 150, 1980.

156. Alvarino, A., The relation between the distribution of zooplankton predators and anchovy larvae., *Rapp. P. -V. Reun. Cons. Int. Explor. Mer,* 178, 197, 1981.

157. Fraser, J. H., The ecology of the ctenophore *Pleurobranchia pileus* in Scottish water, *J. Cons. Cons. Int. Explor. Mer.,* 33, 149, 1970.

158. Lillelund, K. and Lasker, R., Laboratory studies on predation by marine copepods on fish larvae, *Fish. Bull.,* 69, 655, 1971.

159. Theilacker, G. H. and Lasker, R., Laboratory studies of predation by euphausid shrimps on fish larvae, in *The Early Life History of Fish,* Blaxter, J. H. S., Ed., Springer-Verlag, N. Y., 1974, 287.

160. Hunter, J. R. and Kimbrall, C. A., Egg canabalism in the northern anchovy, *Engraulis mordax, Fish. Bull.,* 78, 811, 1980.

161. Lucas, J. R., Feeding ecology of the gulf silverside *Menidia peninsulae* near Crystal River, Florida USA, with notes on its life history, *Estuaries,* 5, 138, 1982.

162. Kulka, D. W. and Stobo, W. T., Winter distribution and feeding of mackerel *Scomber scombrus* on the Scotian Shelf and outer Georges Bank (North Atlantic Ocean) with reference to the winter distribution of other finfish species, *Can. Tech. Rep. Fish. Aquat. Sci.,* 1038, 1, 1981.

163. Grave, H., Food and feeding of mackerel larvae and early juveniles in the North Sea, *Rapp. P. -V. Reun. Cons. Int. Explor. Mer,* 178, 454, 1981.

164. Colin, P. L., Filter feeding and predation on the eggs of *Thallasoma* sp. by the scombrid fish *Rastrelliger kanagurta, Copeia,* 1976, 596, 1976.

165. Hobson, E. S. and Chess, J. R., Trophic relationships among fishes and plankton at Enewetok Atoll, Marshall Islands, *Fish. Bull.,* 76, 133, 1978.

166. Kawai, T. and Isibasi, K., Comparative biology of Japanese fish. 3. A quantitative analysis of larval mortality by cannibalism, *Bull. Tokai Reg. Fish Res. Lab.,* 100, 79, 1979.

167. Koslow, J. A., Feeding selectivity of schools of northern anchovy, *Engraulis mordax,* in the southern California Bight, *Fish. Bull.,* 79, 131, 1981.

168. Cushing, D. H., The predation cycle and numbers of marine fish, *Symp. Zool. Soc. Lond.,* 29, 213, 1972.

169. Houde, E. D., Some recent advances and unsolved problems in the culture of marine fish larvae, *Proc. World Maricult. Soc.,* 3, 83, 1973.

170. Jones, R. and Hall, W. B., Some observations on the population dynamics of the larval stage in the common gadoids, in *The Early Life History of Fish,* Blaxter, J. H. S., Ed., Springer-Verlag, N. Y., 1974, 87.

171. Arthur, D. L., Distribution, size and abundance of microcopepods in the California Current system and their possible influence on survival of marine teleost larvae, *Fish. Bull.,* 75, 601, 1977.

172. Vlymen, W. J., A mathematical model of the relationship between larval anchovy (*Engraulis mordax*) growth, prey microdistribution, and larval behavior. *Environ. Biol. Fishes,* 2, 211, 1977.

173. Lasker, R. and Smith, P. E., Estimation of the effects of environmental variations on eggs and larvae of the northern anchovy, *Calif. Coop. Oceanic Fish. Invest. Rep.,* 19, 128, 1977.

174. Houde, E. D., Critical food concentrations for larvae of three species of subtropical marine fishes, *Bull. Mar. Sci.,* 28, 395, 1978.

175. Houde, E. D. and Schekter, R. C., Feeding by marine fish larvae: developmental and functional responses, *Environ. Biol. Fishes,* 5, 315, 1980.

176. O'Connell, C. P., Percentage of starving northern anchovy, *Engraulis mordax.* as estimated by histological methods, *Fish. Bull.,* 78, 475, 1980.

177. Kondo, K., The recovery of the Japanese sardine: the biological basis of stock size fluctuations, *Rapp. P. -V. Reun. Cons. Int. Explor. Mer,* 177, 322, 1980.

178. Lasker, R., The role of a stable ocean in larval fish survival and subsequent recruitment, in *Marine Fish Larvae: Morphology, Ecology and Relation to Fisheries,* Lasker, R., Ed., Sea Grant Publ. (Washington University), 1981, 80.

179. Lasker, R., Factors contributing to variable recruitment of the northern anchovy (*Engraulis mordax*) in the California Current — contrasting years, 1975 through 1978, *Rapp. P. -V. Reun. Cons. Int. Explor. Mer,* 178, 375, 1981

INDEX

N

O

0 00 02 0420289 9

MIDDLEBURY COLLEGE